Gradient Inequalities

with Applications to Asymptotic Behavior
and Stability of Gradient-like Systems

Mathematical Surveys and Monographs

Volume 126

Gradient Inequalities

with Applications to Asymptotic Behavior and Stability of Gradient-like Systems

Sen-Zhong Huang

American Mathematical Society

2000 *Mathematics Subject Classification.* Primary 35A15, 35Bxx, 35Kxx, 37L15, 47J35; Secondary 35Q80, 47Hxx.

For additional information and updates on this book, visit
www.ams.org/bookpages/surv-126

Library of Congress Cataloging-in-Publication Data

Huang, Sen-Zhong, 1962–
Gradient inequalities with applications to asymptotic behavior and stability of gradient-like systems / Sen-Zhong Huang.
p. cm. — (Mathematical surveys and monographs, ISSN 0076-5376 ; v. 126)
Based on the Habilitationsschrift–University of Rostock, 2003.
Includes bibliographical references and index.
ISBN 0-8218-4070-3 (alk. paper)
1. Topological dynamics. 2. Differential equations—Asymptotic theory. 3. Variational Inequalities (Mathematics) 4. Calculus of variations. I. Title. II. Mathematical surveys and monographs ; no. 126.

QA611.5.H83 2006
515′.64—dc22 — 2005058916

∞ The paper used in this book is acid-free and falls within the guidelines
established to ensure permanence and durability.
Visit the AMS home page at `http://www.ams.org/`

10 9 8 7 6 5 4 3 2 1 11 10 09 08 07 06

Contents

Preface

The term "gradient inequality" appears for the first time in the monograph of J. W. Neuberger [91], where the important role of gradient inequalities in proving convergence of solutions to gradient systems is investigated.

The idea that gradient inequalities force convergence in gradient systems was known to S. Łojasiewicz in the 1960s. Łojasiewicz's celebrated proof [82] of gradient inequalities for analytic gradient maps on finite-dimensional spaces is certainly very complicated, and in fact, the most difficult part in applying the above idea is that of establishing gradient inequalities.

Infinite-dimensional gradient inequalities first appeared in 1983 in the pioneering work of L. Simon [107]. Simon's gradient inequalities are concerned with analytic gradient maps associated with variational problems; his idea of establishing gradient inequalities has been extended by several authors.

This monograph gives a comprehensive study of gradient inequalities and their applications in proving convergence of solutions to gradient-like systems. Many of the results published here are new, and a lot of them are established by the author himself. The very broad applications fields of these results include the mathematical modeling of physical problems as well as calculus of variations, image processing, geometric evolution problems arising from geometric interests and optimization.

This monograph is based on my Habilitationsschrift finished in 2003 at the University of Rostock. In preparing the manuscript I have benefited from a number of friends and colleagues. In particular, I wish to thank Prof. Peter Takáč for his long-time support and encouragement. I am deeply indebted to Prof. Rainer Nagel, Prof. Jerome A. Goldstein and Prof. John W. Neuberger for their continuous encouragement and help both in mathematics and life. I thank Dr. Ralph Chill, Dr. Thomas Elsken, Prof. Jacqueline Fleckinger, Prof. Alain Haraux, Prof. Jan Prüss, Dr. Ian Schindler, Dr. Roland Schnaubelt, Dr. Michael Ulm and Prof. Ulf Schlotterbeck for their valuable comments and discussions. My special thanks go to Dr. Simon Brendle for his kind help in obtaining the monograph of S. Łojasiewicz [82] and for many useful discussions.

The author would like to thank two anonymous referees for their suggestions that have greatly improved the theories presented in this volume.

The author thanks Mr. Edward G. Dunne, Mrs. Christine Thivierge and Mr. Gil Poulin of AMS for their professional help.

I dedicate this book to my wife Yan, and my sons Leo and Simon.

Hamburg, August 2005 Sen-Zhong Huang

CHAPTER 1

Introduction and overview of the results

The asymptotic (long-time) behavior of dynamical systems may be very arbitrary. In recent works [**64, 65**], the author has shown by use of the construction method that the attractors for the two typical models arising from applied sciences, the Navier-Stokes (NS) equations $u_t + (u \cdot \nabla)u = \nu\Delta u - \nabla p + f$, $\operatorname{div} u = 0$ and the Reaction-Diffusion (RD) systems $u_t = D\Delta u + F(u)$ [**118**], may be very complicated if the time-dependent external force f for the (NS) equation (resp. the vector supply term F for the (RD) system) has been applied from some infinite-dimensional and complicated source. In general, attractors are considered as the ensemble of infinitely many trajectories. However, the attractor we have constructed for the (NS) equation (resp. for the (RD) system) is generated by a single trajectory. Moreover, the very complicated attractor for the autonomous (RD) system is a part of the equilibria set of that system.

Although the afore mentioned (RD) system does not belong to the class of so-called *gradient-like* systems, it shares the typical property of a gradient-like system that every relatively compact trajectory asymptotically approaches a set of equilibria [**51**, Lemma 3.8.2, p. 50]. This raises an interesting question as to whether the situation would be better in *gradient systems*. Unfortunately, an example due to P. Poláčik and F. Simondon [**96**] (see also [**95**], [**29**]), by modifying a construction of P. Poláčik and K. P. Rybakowski [**95**] (the latter is based on an example given by J. Palis and W. de Melo [**93**]), gives a C^∞ function f such that the following nonlinear heat equation

$$\text{(1.1)} \qquad \begin{cases} \frac{\partial u}{\partial t} = \Delta u - f(x, u(t,x)), & x \in \Omega, \quad t > 0; \\ u(t,x) = 0 \quad x \in \partial\Omega, & t > 0; \\ u(0,x) = u_0(x), & x \in \Omega, \end{cases}$$

in the bounded domain $\Omega \subset \mathbb{R}^N$ (with $N \geq 2$ and a C^2 boundary $\partial\Omega$), has a bounded solution that asymptotically approaches a continuum of stationary solutions of that equation. Eq. (1.1) can be rewritten in its gradient system form

$$\text{(1.2a)} \qquad \dot{u} = -\mathcal{E}'(u) \quad (\dot{u} \equiv \partial u/\partial t),$$

where $\mathcal{E}'(u) := -\Delta u + f(x,u)$ is the *gradient* of the *energy functional*

$$\text{(1.2b)} \qquad \mathcal{E}(u) := \frac{1}{2}\int_\Omega |\nabla u|^2\,dx + \int_\Omega F(x,u)\,dx \quad (F_u = f).$$

There are two ingredients in the above example of Poláčik and Simondon [**96**]: one is the higher dimension of the domain Ω and the other one is the smoothness of the function f. In fact, if $\Omega = (a, b) \subset \mathbb{R}^1$ is an interval, then convergence to a single equilibrium holds under very general hypotheses on f and the boundary conditions; see T. J. Zelenyak [**124**] and H. Matano [**86**] (also [**53**]). On the other hand, analyticity of the function f helps to overcome the difficulties encountered in higher-dimensional cases. It was L. Simon [**107**] who first established the long-time convergence to one single equilibrium of solutions of Eq. (1.1), for any bounded domain $\Omega \subset \mathbb{R}^N$, under the assumption that the nonlinearity $f(x, u)$ is (real) analytic with respect to u.

Simon's proof of convergence to one single equilibrium is deduced from a **gradient inequality** which compares the *gradient* of $\mathcal{E}$ to an *increment* of $\mathcal{E}$, and it takes the form

$$
\begin{aligned}
&|\mathcal{E}(u) - \mathcal{E}(\varphi)|^\theta \leq \|\Delta u - f(\cdot, u)\|_{L^2(\Omega)} \\
&\forall u \in C^{2,\alpha}(\bar{\Omega}), \|u - \varphi\|_{C^{2,\alpha}} < \sigma,
\end{aligned} \tag{1.3}
$$

where $\theta \in [1/2, 1)$ and $\sigma > 0$ are constants, and $\varphi \in C^{2,\alpha}(\bar{\Omega})$ is a classical solution of the Dirichlet boundary value problem

$$
\begin{cases} -\mathcal{E}'(\varphi) = \Delta\varphi - f(x, \varphi(x)) = 0, & x \in \Omega \\ \varphi(x) = 0, \quad x \in \partial\Omega. \end{cases} \tag{1.4}
$$

In proving the gradient inequality (1.3), Simon used a deep result of S. Łojasiewicz [**81, 82, 83**] (Łojasiewicz's theorem), which asserts that if $\Gamma : \mathbb{R}^N \to \mathbb{R}$ is a function that is (real) analytic in a neighborhood of a given critical point $a \in \mathbb{R}^N$ (i.e., $\nabla\Gamma(a) = 0$), then there exist some constants $\theta \in [1/2, 1)$ and $\sigma > 0$ such that

$$
|\Gamma(x) - \Gamma(a)|^\theta \leq |\nabla\Gamma(x)| \quad \forall x \in \mathbb{R}^N, |x - a| < \sigma. \tag{1.5}
$$

The constants θ and σ depend on the function Γ and the point a.

The above cited work of L. Simon was published in 1983. It seems that it was forgotten for a long time. It is only very recently (1997) that F.-H. Lin and Q. Du [**78**] used the same idea as L. Simon to prove the convergence of solutions to the (autonomous) Ginzburg-Landau (GL) model equations for superconductivity. One of the reasons for this delay may be that the original work of Simon is very difficult to read. This situation changed with the works of A. Haraux and M. A. Jendoubi [**58, 71, 72, 55**]. Haraux and Jendoubi (especially Jendoubi [**72**]) have simplified the original proof procedures and further developed Simon's method. There are several subsequent works in extending the above Łojasiewicz-Simon-type (LS-type) gradient inequalities and deducing convergence results from these gradient inequalities in other systems, e.g., E. Feireisl and F. Simondon [**40**] in porous medium equations, E. Feireisl and P. Takáč [**41**] and S.-Z. Huang and P. Takáč [**70**] in time-dependent (GL) equations, P. Rybka and K.-H. Hoffmann [**101, 102**] in

Cahn-Hilliard equations and equations of viscoelasticity with capillarity but without initial effect, R. Chill, E. Fašangová and J. Prüss [**20**] in Cahn-Hilliard equations with dynamic boundary conditions, P. Takáč [**115**] in (RD) systems of the form (1.1) where the Laplacian has been replaced by a general strongly elliptic operator and the nonlinearity f is also assumed to be analytic in the second argument u but is allowed to have bad growth conditions for the first argument x near the boundary of the domain, A. Haraux and M. A. Jendoubi [**59, 60, 61**], especially on damped semilinear wave equations, S. Aizicovici, E. Feireisl, F. Issard-Roch and H. Petzeltová [**2, 3, 4, 37, 38**] on phase-field systems, the recent work of R. Chill and E. Fašangová [**21**] on semilinear wave equations with dissipation of memory type, and the works of R. Chill [**18**] and R. Chill and M. A. Jendoubi [**21**] on abstract evolutionary equations.

1. The methodology

Before proceeding, we sum up the methodology used by all of the above cited works.

Suppose we are given a C^1 functional

$$\mathcal{E} : U \subset X \to \mathbb{R} \tag{1.6a}$$

on an open subset U of a (real) Banach space X. Then its gradient

$$\mathcal{M} := \mathcal{E}' : U \to X' \tag{1.6b}$$

is a continuous map from U to X'. Suppose that a stronger continuity of the gradient map $\mathcal{M}$ can be gained in the sense that there exists a subspace Y of X', which becomes a Banach space under its own norm $\|\cdot\|_Y$, such that

$$\|y\|_Y \geq \|y\|_{X'} \quad \forall y \in Y, \tag{1.6c}$$

and such that $\mathcal{M}$ maps U into Y, and the map

$$\mathcal{M} = \mathcal{E}' : U \to Y \tag{1.6d}$$

is continuous, i.e., the map $U \ni u \mapsto \mathcal{M}(u) \in Y$ is continuous with respect to the stronger Y norm topology. Consider a given subset $V \subset U$. We say that $\mathcal{M} = \mathcal{E}'$ satisfies a **gradient inequality** in V, with respect to the Y norm, if there exists a function ϕ with $0 \leq \frac{1}{\phi} \in L^1_{loc}(\mathbb{R})$ (in notation, $\phi \in \mathcal{G}$) such that

$$\phi(\mathcal{E}(u)) \leq \|\mathcal{E}'(u)\|_Y \quad \forall u \in V. \tag{GI}$$

Given a point $\varphi \in U$, we say that $\mathcal{M} = \mathcal{E}'$ satisfies a gradient inequality near φ if it satisfies a gradient inequality in some neighborhood of φ.

Clearly, $\mathcal{M} = \mathcal{E}'$ satisfies a gradient inequality near every regular point of $\mathcal{E}$. Hence, a gradient inequality is a matter concerning the behavior of $\mathcal{E}$ near its critical points.

Two more remarks should be taken into consideration.

First, we note that the norms of $\mathcal{E}'(u)$ are not evaluated in the X' norm but in the bigger Y norm. This follows from an easy observation: the larger the norms on the right-hand side of the inequality (GI) being evaluated, the better the chance of the existence of a function $\phi \in \mathcal{G}$ satisfying that inequality is. However, for the purpose of deducing convergence results from gradient inequalities, one has to try to estimate the norms in a relatively large subspace of X', because such a convergence occurs in general only with respect to a topology very similar to the topology induced by the norm in X'. This is a crucial point in establishing gradient inequalities.

Second, the condition that $\frac{1}{\phi}$ must be locally integrable is essentially important for applications in proving convergence to one single equilibrium of trajectories. To explain this point, we review the following results in Hilbert spaces, the proofs of which illustrate the methodology used originally by L. Simon and later by others and which will be developed in our present work.

Let H be a Hilbert space, and let $\mathcal{E} : X \to \mathbb{R}$ be a functional that is defined in a subspace X of H and has a gradient $\mathcal{E}'$ with $\mathcal{E}'(X) \subset H$ such that the map $X \ni u \mapsto \mathcal{E}'(u) \in H$ is continuous. We consider the following gradient system in H :

$$\dot{u}(t) = -\mathcal{E}'(u(t)), \quad t \geq 0. \tag{1.7}$$

The general working schema of L. Simon and others can be sketched as follows: Assuming that u is a global solution of (1.7), which is compact in the sense that its orbit $O(u) := \{u(t) : t \geq 0\}$ is contained in some compact subset of X, the goal is to establish the long-time convergence of u to some single point in X. To do so, it suffices to verify the weaker convergence that $\|u(t) - \varphi\|_H \to 0$ as $t \to \infty$ for some $\varphi \in H$. The reason is that the convergence $\|u(t) - \varphi\|_H \to 0$ implies that if $\tilde{\varphi} \in X$ is any cluster point of $O(u)$ with respect to the X norm, then we must also have $\|u(t) - \tilde{\varphi}\|_H \to 0$ as $t \to \infty$ and thus $\tilde{\varphi} = \varphi$. Therefore, $\varphi \in X$ and φ is the unique cluster point of the precompact trajectory u; thus we must have $u(t) \to \varphi$ in X as $t \to \infty$.

One condition that implies the convergence of u in H is the following:

$$I(u) := \int_0^\infty \|\dot{u}(t)\|_H \, dt < \infty.$$

This integral $I(u)$ can be thought of as the *length* of the curve $u : [0, \infty) \to X \subset H$ evaluated in the H norm. In all of the afore mentioned works, the convergence of u in the H norm topology is exactly this kind.

Our first result (Proposition 4.1 of Chapter 3) shows that the finiteness $I(u) < \infty$ is equivalent to the validity of a gradient inequality in the orbit $O(u)$, or equivalent to an appropriate convergence rate of the energies.

THEOREM 1.1. *Assume that $u : \mathbb{R}_+ \to X$ is a classical solution of (1.7) such that*

$$\sup\{|\mathcal{E}(u(t))| : t \geq 0\} < \infty.$$

Then the following assertions **(i)-(iii)** *are equivalent:*

(i) *$\mathcal{E}'$ satisfies a gradient inequality in the orbit $O(u) := \{u(t) : t \geq 0\}$ with respect to the H norm; i.e., there holds the inequality*

$$\phi(\mathcal{E}(v)) \leq \|\mathcal{E}'(v)\|_H \quad \forall v \in O(u) \tag{1.8}$$

with some function ϕ such that $0 \leq \frac{1}{\phi} \in L^1_{loc}(\mathbb{R})$.

(ii) *There holds*

$$I(u) = \int_0^\infty \|\dot{u}(t)\|_H \, dt < \infty. \tag{1.9}$$

(iii) *There exists some function $\tilde{\phi}$ with $0 \leq \frac{1}{\tilde{\phi}} \in L^1_{loc}(\mathbb{R})$ such that*

$$\int_0^\infty \tilde{\phi}(\mathcal{E}(u(t))) \, dt < \infty. \tag{1.10}$$

We add some comments to these conditions. Note that there holds

$$\frac{d}{dt}\mathcal{E}(u(t)) = -\|\mathcal{E}'(u(t))\|_H^2 = -\|\dot{u}(t)\|_H^2 \leq 0. \tag{$*$}$$

Therefore, the energy function $\mathcal{E}(u(t))$ is nonincreasing, and thus the given boundedness assumption $\sup\{|\mathcal{E}(u(t))| : t \geq 0\} < \infty$ implies that the limit $a := \lim_{t\to\infty} \mathcal{E}(u(t))$ exists and is finite. From this we see that we can always find a nonnegative function $\tilde{\phi}(x)$ whose decay to zero as $x \to a$ is so rapid that the convergence in (1.10) occurs.

The crucial requirement for such a function $\tilde{\phi}$ is the extra local integrability $0 \leq 1/\tilde{\phi} \in L^1_{loc}(\mathbb{R})$ such that the decay $\tilde{\phi}(x) \to 0$ as $x \to a$ is not allowed to be too fast. Therefore, the condition in **(iii)** will be satisfied only if the energy function $\mathcal{E}(u(t))$ converges fast enough as $t \to \infty$.

On the other hand, by integrating $(*)$ we obtain that

$$\mathcal{E}(u(t)) - a = \int_t^\infty \|\mathcal{E}'(u(s))\|_H^2 \, ds = \int_t^\infty \|\dot{u}(s)\|_H^2 \, ds \tag{$**$}$$

for all $t \geq 0$. From $(**)$ we see that the requirement that the energy function $\mathcal{E}(u(t))$ converges fast enough as $t \to \infty$ is equivalent to the requirement that the norm function $\|\mathcal{E}'(u(t))\|_H = \|\dot{u}(t)\|_H$ $(t \geq 0)$ decays to zero fast enough as $t \to \infty$. The implication **(ii)$\Longrightarrow$(iii)** states that this is the case if $\int_0^\infty \|\dot{u}(t)\|_H \, dt < \infty$, while the other implication **(i)$\Longrightarrow$(iii)** says that this is also the case if the decay $\|\mathcal{E}'(u(t))\|_H \to 0$ as $t \to \infty$ dominates the decay $\mathcal{E}(u(t)) \to a$ as $t \to \infty$ in such a way that the gradient inequality (1.8) holds true for some $\phi \in \mathcal{G}$.

Now we turn to the convergence implications of gradient inequalities. As previously shown, the convergence $I(u) < \infty$ in Theorem 1.1-**(ii)** implies the usual convergence $\|u(t) - \varphi\|_H \to 0$ as $t \to \infty$ for some $\varphi \in H$. If in addition we know that the orbit $O(u)$ is contained in a compact subset of X, then we have the stronger convergence $\|u(t) - \varphi\|_X \to 0$ as $t \to \infty$ and $\mathcal{E}'(\varphi) = 0$. The following Theorem 1.2 (a special case of Theorem 3.3 of Chapter 3) has extended the above result in the sense that the validity of a gradient inequality in the orbit $O(u)$ can be replaced by the validity of a gradient inequality in the neighborhood of some cluster point of the orbit.

THEOREM 1.2. *Assume that* $u : \mathbb{R}_+ \to X$ *is a classical solution of (1.7) such that its orbit* $O(u)$ *is contained in a compact subset of* X *and has a cluster point* φ *near which* $\mathcal{E}'$ *satisfies a gradient inequality (with respect to the* H *norm), i.e., there exist a positive constant* $\sigma > 0$ *and a function* ϕ *with* $0 \le \frac{1}{\phi} \in L^1_{loc}(\mathbb{R})$ *such that*

$$\phi(\mathcal{E}(v)) \le \|\mathcal{E}'(v)\|_H \quad \forall v \in X, \|v - \varphi\|_X < \sigma. \tag{1.11}$$

Then $u(t)$ *converges as* $t \to \infty$ *to* φ *with respect to the* X *norm, and* φ *is an equilibrium of (1.7). Moreover,* $I(u) = \int_0^\infty \|\dot{u}(t)\|_H \, dt < \infty$.

Sketch of the proof of Theorem 1.2. Let

$$H(t) := \mathcal{E}(u(t)), \quad t \ge 0.$$

We have

$$-\dot{H}(t) = \langle \mathcal{E}'(u(t)), -\dot{u}(t) \rangle = \|\mathcal{E}'(u(t))\|_H \cdot \|\dot{u}(t)\|_H \tag{1.12a}$$

for all $t \ge 0$. This implies that $H(t)$ is nonincreasing and is bounded by the precompactness of the orbit $O(u)$. Let $a := \lim_{t\to\infty} H(t)$ and consider the function

$$\Phi(x) := \int_a^x \frac{1}{\phi(s)} \, ds \quad (x \in \mathbb{R}).$$

Φ is absolutely continuous, since $\frac{1}{\phi}$ is locally integrable (this explains the main point of why we must assume the local integrability of $\frac{1}{\phi}$). Moreover, we have, for the composition $G(t) := \Phi(H(t))$ $(t \ge 0)$, that

$$-\dot{G}(t) = -\dot{H}(t)/\phi(H(t)) \quad \text{a.e.} \quad t \ge 0. \tag{1.12b}$$

We choose an increasing unbounded sequence $\{t_n\}_{n\ge1} \subset \mathbb{R}_+$ such that $\|u(t_n) - \varphi\|_X \to 0$ as $n \to \infty$.

Let $\varepsilon \in (0, \sigma)$, where $\sigma > 0$ is the number involved in the gradient inequality (1.11). Choose $N(\varepsilon) \in \mathbb{N}$ to be so large that $\|u(t_n) - \varphi\|_X < \varepsilon$ for all $n \ge N(\varepsilon)$. For each $n \ge N(\varepsilon)$, we let

$$\tilde{t}_n := \sup\{t \ge t_n : \quad \|u(s) - \varphi\|_X < \varepsilon \quad \forall t_n \le s < t\}.$$

(Intuitively, $\tilde{t}_n$ is the first execution time during which the section $\{u(t) : t \ge t_n\}$ escapes from the neighborhood $\{v \in X : \|v - \varphi\|_X < \varepsilon\}$.)

We claim:

(C) For each $\varepsilon \in (0, \sigma)$ there exists an $N \ge N(\varepsilon)$ such that $\tilde{t}_N = \infty$, i.e.,

$$\|u(t) - \varphi\|_X < \varepsilon \quad \forall t \ge t_N. \tag{1.13}$$

Note that **(C)** implies the convergence $\lim_{t\to\infty} \|u(t) - \varphi\|_X = 0$ as well as the validity of the gradient inequality (1.11) in the section $\{u(t) : t \ge t_N\}$.

To establish **(C)**, we suppose, in contrast, that for some $\varepsilon_0 \in (0, \sigma)$ we can find for each $n \geq N_0 := N(\varepsilon_0)$ an s_n, $s_n > t_n$, such that

$$\|u(t) - \varphi\|_X < \varepsilon_0 \quad \forall t \in [t_n, s_n) \tag{1.14a}$$

but

$$\|u(s_n) - \varphi\|_X \geq \varepsilon_0. \tag{1.14b}$$

Fix $n, n \geq N_0$. For $t \in [t_n, s_n)$ we have $\|u(t) - \varphi\|_X < \varepsilon_0 < \sigma$. Therefore, the gradient inequality (1.11) applies to each $v = u(t)$ with $t \in [t_n, s_n)$ and yields

$$\phi(\mathcal{E}(u(t))) \leq \|\mathcal{E}'(u(t))\|_H \quad \forall t \in [t_n, s_n). \tag{1.15}$$

Using (1.12b) and (1.12a), (1.15) implies that

$$-\dot{G}(t) \geq \|\dot{u}(t)\|_H \quad \text{a.e.} \quad t \in [t_n, s_n) \tag{1.16}$$

and thus, by integration,

$$\int_{t_n}^{s_n} \|\dot{u}(t)\|_H \, dt \leq G(t_n) - G(s_n).$$

Consequently,

$$\begin{aligned} \|u(s_n) - \varphi\|_H &\leq \|u(t_n) - \varphi\|_H + \int_{t_n}^{s_n} \|\dot{u}(t)\|_H \, dt \\ &\leq \|u(t_n) - \varphi\|_H + (G(t_n) - G(s_n)) \to 0 \end{aligned}$$

as $n \to \infty$. This implies that if $\tilde{\varphi}$ is any cluster point of the precompact sequence $\{u(s_n) : n \geq N_0\}$ in X, then we must have $\|\tilde{\varphi} - \varphi\|_H = 0$ and thus $\tilde{\varphi} = \varphi$. Hence, φ is the unique cluster point of the precompact sequence $\{u(s_n) : n \geq N_0\}$ in X and thus $\|u(s_n) - \varphi\|_X \to 0$ as $n \to \infty$, contradicting (1.14b).

In conclusion, we have established **(C)**. As result, it implies that (1.15) holds true for all sufficiently large t. Again, this yields by integration the finiteness $I(u) < \infty$. Clearly, $\mathcal{E}'(\varphi) = 0$. □

REMARK 1.3. The above also shows the following attraction property of an open subset $U_1 \subset X$ in which the gradient map $\mathcal{E}' : X \to H$ satisfies a gradient inequality: If a trajectory $u(t)$ of the gradient system (1.7) visits the neighborhood U_1 infinitely many times as $t \to \infty$, then the trajectory $u(t)$ will eventually be attracted to that neighborhood U_1.

As seen in Theorem 1.2, the validity of a gradient inequality near some cluster point of a precompact trajectory suffices to ensure the convergence to one single equilibrium of that trajectory. Thus, the preparing of suitable gradient inequalities is in general the first task before proving the convergence.

The idea of establishing gradient inequalities, which will be presented in the next chapter, can be sketched as follows: First, the involved gradient map must be a Fredholm map of index zero. This Fredholm property will be used (by the Lyapunov-Schmidt reduction) to reduce the involved infinite-dimensional problem to a corresponding finite-dimensional problem. Second, an expected infinite-dimensional gradient inequality will be deduced, provided that there holds an appropriate gradient inequality for the corresponding finite-dimensional problem. The best known and universal finite-dimensional gradient inequality is the above cited Łojasiewicz gradient inequality (1.5) for *analytic* functions. Because of this, we are able to show (in Chapter 2) that many analytic gradient maps satisfy a gradient inequality of Łojasiewicz-Simon type.

Let us review the gradient inequality (1.3) of L. Simon. It is not surprising, as compared to Łojasiewicz's inequality (1.5), that the function ϕ has the form $\phi_\theta(s) := |s - \mathcal{E}(\varphi)|^\theta$ for all $s \in \mathbb{R}$. The requirement $\theta < 1$ for the exponent corresponds exactly to the local integrability of $1/\phi_\theta$. It is also of little surprised why Simon chose $X = C^{2,\alpha}(\bar{\Omega})$. This choice makes the functional $\mathcal{E}$ an analytic functional from X to $\mathbb{R}$ and such that the gradient $\mathcal{E}'$ has a range contained in $C^\alpha(\bar{\Omega})$. Thus, any Banach space Y between $C^\alpha(\bar{\Omega})$ and the dual of $C^{2,\alpha}(\bar{\Omega})$ can be considered as a possible candidate for establishing the gradient inequality (GI) for this case. In the original proof of Simon, the gradient inequality with the choice $Y = C^\alpha(\bar{\Omega})$ is fairly easy to make. However, jumping from the space $C^\alpha(\bar{\Omega})$ to the Hilbert space $Y = L^2(\Omega)$ is certainly a subtle skill. What Simon needs for this jump is the classical regularity results shared by the Laplacians; this regularity property corresponds exactly to the Fredholm property of the semilinear elliptic operators. As seen in Theorem 1.2, the validity of a gradient inequality with respect to the $H = L^2(\Omega)$ norm suffices to guarantee the convergence to one single equilibrium for compact solutions of the nonlinear heat equation (1.1).

The remainder context of this chapter is concerned with the sketch of the convergence results deduced from gradient inequalities and applications of these convergence results, which we will elaborate on in the latter chapters. Each of these results is presented in both its abstract form and in its more concrete form for systems settled in Hilbert spaces. Moreover, we will give a brief synopsis for further studies in these topics.

2. Convergence results for gradient-like trajectories

Our setting is as follows. Let U be an open subset of a Banach space X and $\mathcal{E}' : U \to \mathbb{R}$ a C^1 functional. We assume further that there is a subspace Y of the dual X' such that $\mathcal{E}'(U) \subset Y$ and that the map $U \ni u \mapsto \mathcal{E}'(u) \in Y$ is continuous.

By a *trajectory* in U, we mean a continuous map $u : [0, T) \to U$ that has a weak derivative $\dot{u}$ taking values in Y'. Our study is concentrated in these *gradient-like trajectories* $u : [0, \infty) \to U$ that satisfy a growth condition of the form:

$$\begin{cases} \langle -\mathcal{E}'(u(t)), \dot{u}(t) \rangle_{Y \times Y'} \geq \|\mathcal{E}'(u(t))\|_Y \cdot \|\gamma \dot{u}(t)\|_{Y'} + F'(t), \\ \|\mathcal{E}'(u(t))\|_Y \geq \|\gamma \dot{u}(t)\|_{Y'} + G'(t) \end{cases} \tag{1.17a}$$

for all $t \geq 0$, where $\gamma : Y' \to Y'$ is an **injective** and continuous map (which could be nonlinear), and $F, G : \mathbb{R}_+ \to \mathbb{R}_+$ are nonincreasing C^1 functions satisfying

$$\lim_{t \to +\infty} (F(t) + G(t)) = 0 \quad \text{and} \quad \int_0^\infty \sqrt{F(t)}\, dt < \infty. \tag{1.17b}$$

The map γ as well as the functions F, G may depend on the given trajectory. A simplified version of (1.17a,b) is obtained by switching F, G into zero and weakening the inequality in the second line of (1.17a):

$$\begin{cases} \langle -\mathcal{E}'(u(t)), \dot{u}(t) \rangle_{Y \times Y'} \geq \|\mathcal{E}'(u(t))\|_Y \cdot \|\gamma \dot{u}(t)\|_{Y'}, \\ \mathcal{E}'(u(t)) = 0 \Longrightarrow \dot{u}(t) = 0 \end{cases} \quad \forall t \geq 0. \tag{1.17$'$}$$

Observe that in (1.17a) and (1.17′), the norms for the gradient $\mathcal{E}'$ are not evaluated in the smaller X' norm but in the larger Y norm. The use of the larger Y norm is due to the fact that, in general, one can only have a gradient inequality in the larger Y norm rather than in the smaller X' norm.

Also note that the solutions considered in Theorem 1.2 satisfy (1.17′) under the choice of the space $Y = H \cong H' = Y'$ and $\gamma =$ identity operator on H. Therefore, Theorem 2.1 (corresponding to Theorem 3.3 of Chapter 3) extends Theorem 1.2.

THEOREM 2.1. *Let $u : [0, \infty) \to U$ be a trajectory satisfying (1.17′). Assume that the orbit $O(u) = \{u(t) : t \geq 0\}$ is precompact in X and has a cluster point $\varphi \in X$ near which the gradient map $\mathcal{E}' : U \to Y$ satisfies a gradient inequality. Then the trajectory u converges as $t \to \infty$ to φ in the X norm topology.*

For Theorem 2.2 (corresponding to Theorem 3.6 of Chapter 3), we need a new notion. For a constant $k \geq 1$, we denote by $\mathcal{G}_k$ the set of all functions $\phi \in \mathcal{G}$ (i.e., $0 \leq \frac{1}{\phi} \in L^1_{loc}(\mathbb{R})$) satisfying the following extra growth condition

$$\phi(x + y) \leq k\phi(x) + |y|^{1/2} \quad \forall x, y \in \mathbb{R}. \tag{1.18}$$

We say that a gradient inequality (GI) is of type $\mathcal{G}_k$ if the corresponding function ϕ involved in (GI) belongs to the class $\mathcal{G}_k$.

THEOREM 2.2. *Let $u : [0, \infty) \to U$ be a trajectory satisfying (1.17a) and (1.17b). Assume that the orbit $O(u) = \{u(t) : t \geq 0\}$ is precompact in X and has a cluster point $\varphi \in X$ near which the gradient map $\mathcal{E}' : U \to Y$ satisfies a gradient inequality of type $\mathcal{G}_k$ with some $k \geq 1$. Then the trajectory u converges as $t \to \infty$ to φ in the X norm topology.*

Comments. (a) We note that the functions

$$\tilde{\phi}_\theta(s) := \frac{1}{2}\min\{|s-a|^\theta, |s-a|^{1/2}\} \quad (s \in \mathbb{R})$$

with constants $a \in \mathbb{R}$ and $\theta \in [1/2, 1)$ belong to the class $\mathcal{G}_2$. Therefore, if we replace the function $\phi_\theta(s) = |s - \mathcal{E}(\varphi)|^\theta$ used in the gradient inequality (1.3) of L. Simon by the smaller $\tilde{\phi}_\theta$ with $a = \mathcal{E}(\varphi)$, then we have a gradient inequality of type $\mathcal{G}_2$. As it will be shown later, a gradient inequality of the type (1.3) will be shared by many other analytic gradient maps and thus Theorem 2.2 can be applied in a wide variety of applications.

(b) Observe that in (1.17a) as well as in (1.17′) we have included a continuous map $\gamma : Y' \to Y'$ that is only required to be injective. Hence the function $\|\gamma\dot{u}(t)\|_{Y'}$, compared to the usual norm function $\|\dot{u}(t)\|_{Y'}$, may have a very quick decay. It is exactly this point that opens up enormous possibilities for applying Theorem 2.2.

3. Applications to gradient-like systems in Hilbert spaces

Results presented below will be proved in the latter chapters, especially in §4 of Chapter 3 for the general case and in Chapter 4 for the special semilinear case.

3.1. General systems. Our setting is as follows.

- H is a Hilbert space and $A : D(A) \subset H \to H$ is a linear self-adjoint and positive definite operator.
- Both X and $\tilde{X}$ are Banach spaces under their respective norms and such that the embeddings

$$X \hookrightarrow H_A \equiv (D(A), \langle\cdot,\cdot\rangle_A), \quad \tilde{X} \hookrightarrow H$$

are continuous. Moreover, $X \subset \tilde{X}$.

- $\mathcal{M} = \mathcal{E}' : X \to \tilde{X}$ is an **analytic** gradient map with the following properties: $\mathcal{M}$ is a Fredholm map of index zero; i.e., for each $u \in X$ the linear bounded operator $\mathcal{M}'(u) \in L(X, \tilde{X})$ is a Fredholm operator of index zero. Moreover, for each fixed $u \in X$ the bounded linear symmetric operator $\mathcal{M}'(u) : X \to \tilde{X}$ has an extension $\mathcal{M}_1(u) : H_A \to H$, which is also a symmetric Fredholm operator of index zero, and such that the map

$$X \ni u \mapsto \mathcal{M}_1(u)A^{-1} \in L(H)$$

is continuous.

THEOREM 3.1. **(Gradient inequalities; Theorem 4.2 of Chapter 2)** *Under the above circumstances, there hold the following assertions:*

(i) The analytic gradient map $\mathcal{E}' : X \to \tilde{X}$ satisfies a gradient inequality of LS-type near every critical point; i.e., if $\varphi \in X$ is a critical point of $\mathcal{E}$, $\mathcal{E}'(\varphi) = 0$, then there exist positive constants c, σ and $\theta \in [1/2, 1)$ such that

$$c \cdot |\mathcal{E}(u) - \mathcal{E}(\varphi)|^\theta \le \|\mathcal{E}'(u)\|_H \quad \forall u \in X, \|u-\varphi\|_X < \sigma. \tag{1.19a}$$

(ii) The analytic gradient map $\mathcal{E}' : X \to \tilde{X}$ satisfies a gradient inequality of type $\mathcal{G}_2$ in every compact subset of X; i.e., for every compact subset $K \subset X$ there exists some $\phi \in \mathcal{G}_2$ such that

$$\phi(\mathcal{E}(u)) \leq \|\mathcal{E}'(u)\|_H \quad \forall u \in K. \tag{1.19b}$$

Our first convergence result is concerned with the following first order gradient-like system in the Hilbert space H:

$$\begin{cases} \dot{u} = -\mathcal{N}(u) + g(t, u), \quad t \geq 0, \\ u \equiv u(t) \in H, \dot{u} \equiv \dot{u}(t) \in H, \end{cases} \tag{1.20a}$$

where $\mathcal{N} : X \to H$ is a continuous map for which there exist a **strictly positive** continuous function $\eta : X \to (0, 1)$ and an **injective** linear and bounded operator $\tilde{J} : H \to H$ such that

$$\begin{cases} \langle \mathcal{E}'(u), \mathcal{N}(u) \rangle_{H \times H} \geq \eta(u) \cdot \|\mathcal{E}'(u)\|_H^2 \\ \|\tilde{J}\mathcal{N}(u)\|_H \leq (1/\eta(u)) \cdot \|\mathcal{E}'(u)\|_H \end{cases} \qquad \forall u \in X. \tag{1.20b}$$

Moreover, the time-dependent perturbation $g : \mathbb{R}_+ \times X \to H$ is a continuous map satisfying

$$\begin{cases} \|g(t, u)\|_H \leq g(t) \quad \forall t \geq 0, u \in X; \\ \displaystyle\int_0^\infty \rho(t) g(t)^2 \, dt + \int_0^\infty \frac{1}{\rho(t)} < \infty \\ \text{with some increasing positive function} \quad \rho. \end{cases} \tag{1.20c}$$

Roughly speaking, condition (1.20b) states that $\mathcal{N}$ is a gradient-like map associated to the gradient map $\mathcal{E}'$, while (1.20c) is a condition implying that system (1.20a) will become asymptotically autonomous in a collective way. It is worth noting that condition (1.20c) does not imply the decay $g(t) \to 0$ as $t \to \infty$. In fact, there are functions that satisfy (1.20c), but they have arbitrarily large bounds in a suitable sequence of small intervals.

THEOREM 3.2. **(Theorem 4.2 of Chapter 3)**
Under the above situations we let $u : [0, \infty) \to H$ be a solution of (1.20a) such that the orbit $O(u)$ is contained in a compact subset of the Banach space X. Then $u(t)$ converges as $t \to \infty$ to some $\varphi \in X$ in the X norm topology and $\mathcal{E}'(\varphi) = 0$.

The tool for proving Theorem 3.2 is the gradient inequality (1.19a) or (1.19b).

There are also analogous convergence results (Theorem 4.4 and Theorem 4.5 of Chapter 3) for the following second-order gradient-like systems in the Hilbert space H :

$$\begin{cases} \ddot{u} + \kappa B(t, u, \dot{u}) = \pm \mathcal{E}'(u) + g(t, u, \dot{u}), \quad t \geq 0, \\ u \equiv u(t) \in H, \dot{u} \equiv \dot{u}(t) \in H, \ddot{u} \equiv \ddot{u}(t) \in H, \end{cases} \tag{1.21}$$

where the constant κ is either equal to -1 or 1, and $B : \mathbb{R}_+ \times X \times H \to H$ and $g : \mathbb{R}_+ \times X \times H \to H$ are continuous maps satisfying appropriate coerciveness and boundedness conditions. The choice $\kappa = 1$ (resp. $\kappa = -1$) corresponds to the case where the term $\kappa B(t, u, \dot{u})$ works as a structural damping (resp. a structural excitation) of the system (1.21).

If $\mathcal{E}'(u) = -\Delta u + f(x, u)$ is a semilinear elliptic operator, then (1.21) becomes $\ddot{u}(t) + \kappa B(t, u, \dot{u}) = \pm(-\Delta u + f(x, u)) + g(t, u, \dot{u})$. Under the choice of the negative sign, it is simply the *semilinear wave equation*; while under the choice of the positive sign, it is an equation of *elliptic type*.

We make several remarks.

(a) Theorem 3.2 and Theorems 4.4 and 4.5 of Chapter 3 apply particularly to the semilinear case where $\mathcal{E}'(u) = Au + f(x, u)$ and to the variational case where the functional $\mathcal{E}$ has the form $\mathcal{E}(u) = \int_\Omega F(x, u, Du)\, dx$ with a Carathéodory function $F(x, u, \xi)$ which is uniformly analytic in the u variable and strictly convex in the ξ variable.

(b) Evolutionary equations corresponding to variational functionals are, in general, fully nonlinear. Because of the lack of a powerful theory about the existence of global solutions to such fully nonlinear equations, our results presented in §5 of Chapter 3 for the variational case are somewhat restrictive.

(c) In contrast to the variational case, our results presented in Chapter 4 for the semilinear case are very fruitful. The reason is that there is a complete and powerful theory [**63, 94**] for the existence of global solutions to the corresponding semilinear Cauchy problems. We will sketch some of them below.

(d) Since we have also proved in Theorem 4.3 of Chapter 2 the validity of gradient inequalities of LS-type for some special class of **non-analytic** gradient maps, there are also convergence results analogous to those mentioned above for some special non-analytic cases.

(e) Theorem 3.2 applies to the following special choice of the map $\mathcal{N} = J\mathcal{E}'$; i.e., it applies to the following systems

$$\dot{u} = -J\mathcal{E}'(u) + g(t, u), \tag{1.22}$$

where the operator

$$J : D(J) \subset H \to H$$

is linear, closed, surjective and coercive in the sense that

$$\langle Jy, y\rangle_{H\times H} \geq c \cdot \|y\|_H^2 \quad (y \in D(J)).$$

Typical examples of systems (1.22) are the classical Cahn-Hilliard equations $\dot{u} = \Delta(-\Delta u + f(x, u))$ corresponding to the choices $J = -\Delta$ and $\mathcal{E}'(u) = -\Delta u + f(x, u)$.

(f) Note that the operator J involved in (1.22) is allowed to be **nonlinear** and in some sense to be very arbitrary.

(g) Theorem 3.2 applies particularly to perturbations that satisfy the condition

$$\int_0^\infty t|\log t|^\tau g(t)^2\,dt < \infty \quad \text{with some} \quad \tau > 1.$$

It seems that convergence results for perturbations satisfying the above simple growth condition are sufficient for many application situations.

3.2. Semilinear systems. Below, $(\Omega, \mathcal{B}, \mu)$ denotes a measure space of finite total measure $|\Omega| \equiv \mu(\Omega) < \infty$ and d is a positive integer. We will use $L^p(\Omega;\mathbb{R}^d)$ to denote the L^p spaces of vector-valued functions of d components over the given measure space Ω.

Our study focuses on the following first-order semilinear systems

$$\begin{cases} \dot{u} = -\mathcal{E}_F'(u) + g(t,u), \\ \mathcal{E}_F'(u) \equiv Au + f(x,u), \quad f \equiv \nabla_u F \end{cases} \tag{1.23a}$$

over the Hilbert space

$$\mathcal{H} := L^2(\Omega;\mathbb{R}^d), \tag{1.23b}$$

where $\mathcal{E}_F$ is a functional of the form

$$\mathcal{E}_F(u) := \frac{1}{2}\|Bu\|_{L^2}^2 + \int_\Omega F(x,u(x))\,d\mu(x) \quad (B := \sqrt{A}) \tag{1.23c}$$

for which the linear part A and the nonlinearity F satisfy the following conditions (i)-(ii):

(i) **(Compactness and submarkovian property).** The linear operator

$$A = \operatorname{diag}(A_1, ..., A_d) : D(A) \subset L^2(\Omega;\mathbb{R}^d) \to L^2(\Omega;\mathbb{R}^d) \tag{1.24a}$$

is a linear, diagonal, positive semidefinite and symmetric submarkovian generator such that the inverse

$$(1+A)^{-1} : L^2(\Omega;\mathbb{R}^d) \to L^2(\Omega;\mathbb{R}^d) \tag{1.24b}$$

is compact. Moreover, there exist $\beta \in (0,1)$ and $q \in [2,\infty)$ such that the embedding

$$D(A^\beta \restriction L^q) \cap L^\infty(\Omega;\mathbb{R}^d) \hookrightarrow L^\infty(\Omega;\mathbb{R}^d) \tag{1.24c}$$

is compact.

(ii) **(Analyticity and growth condition).** $F : \Omega \times \mathbb{R}^d \to \mathbb{R}$ is a Carathéodory function that is uniformly *analytic* with respect to the second argument and its gradient $(f_1, ..., f_d) \equiv f := \nabla_u F$ satisfies the following growth condition: There exist three constant vectors $M = (M_1, ..., M_d) \geq 0$, $\tau_+ = (\tau_+^1, ..., \tau_+^d) \geq 0$ and $\tau_- = (\tau_-^1, ..., \tau_-^d) \leq 0$ such that, for all boundary points $u \equiv (u_1, ..., u_d) \in \partial Q$ of the "rectangle" $Q := \{u \in \mathbb{R}^d : \tau_- \leq u \leq \tau_+\}$, there holds

$$\begin{cases} f_j(\cdot,u) \geq M_j & \text{if} \quad u_j = \tau_+^j; \\ f_j(\cdot,u) \leq -M_j & \text{if} \quad u_j = \tau_-^j; \end{cases} \qquad j = 1, 2, ..., d. \tag{1.25}$$

Particularly, for the scalar case $d = 1$, condition (1.25) is equivalent to the following very transparent one:

$$f(\cdot, \tau_+) \geq M, \quad f(\cdot, \tau_-) \leq -M. \tag{1.25$'$}$$

It is very important to note that the growth condition (1.25) (cf. (1.25′) for the scalar case) does not imply, in general, the coercivity of the functional $\mathcal{E}_F$. In fact, it is not difficult to construct functions F that satisfy the above growth condition (1.25′), but the corresponding functionals $\mathcal{E}_F$ are not bounded from below.

Our final condition is on the perturbation

$$g : \mathbb{R}_+ \times L^\infty(\Omega; \mathbb{R}^d) \to L^\infty(\Omega; \mathbb{R}^d).$$

We assume that g is a locally Lipschitz map satisfying

$$\begin{cases} \|g(t,u)\|_{L^\infty} \leq M, \ \|g(t,u)\|_{L^2} \leq g(t) \quad \forall (t,u) \in \mathbb{R}_+ \times L^\infty; \\ \displaystyle\int_0^\infty \rho(t) g(t)^2 \, dt + \int_0^\infty \frac{1}{\rho(t)} < \infty \\ \text{with some increasing positive function} \quad \rho. \end{cases} \tag{1.26}$$

We have the following convergence and stability results.

THEOREM 3.3. **(Convergence; Theorem 1.8 of Chapter 4)** *Under the above circumstances we set* $X := D(\sqrt{A}) \cap L^\infty(\Omega; \mathbb{R}^d)$ *and*

$$X_\tau := \{\varphi \in X : \tau_- \leq \varphi \leq \tau_+\}.$$

Then for each initial value $u_0 \in L^\infty$ *with*

$$\tau_- \leq u_0 \leq \tau_+,$$

the Cauchy problem corresponding to (1.23a) has a unique classical solution

$$u \in C([0,\infty); L^2) \cap C((0,\infty); D(A)) \cap C^1((0,\infty); L^2)$$

for which there exists an equilibrium $\varphi \in X_\tau$ *such that*

$$\lim_{t\to\infty} \|u(t) - \varphi\|_X = 0.$$

Moreover, the part of equilibria

$$\mathcal{S}_\tau := \{\varphi \in X_\tau : \mathcal{E}_F'(\varphi) = 0\}$$

is a compact subset of X.

THEOREM 3.4. **(Stability of ground state; Theorem 1.9 of Chapter 4)** *Under the above circumstances there exists some* $\psi \in \mathcal{S}_\tau$ *such that*

$$\mathcal{E}_F(\psi) \leq \mathcal{E}_F(u) \quad \forall u \in X_\tau.$$

Moreover, this local ground state ψ*, as a stationary solution of the gradient system* $\dot{u} = -\mathcal{E}_F'(u)$*, is Lyapunov stable, and it is uniformly asymptotically stable if it is an isolated stationary solution.*

Both Theorems 3.3 and 3.4 apply to the case where A is a *Dirichlet form operator* [**84**], and especially to the case where $-A$ are Laplacian operators. Related results about semilinear evolutionary problems around the Laplacian operators will be presented in §3, §4 and §5 of Chapter 4. There we will show the convergence and stability of the following model equations arising from applied sciences: the Cahn-Hilliard equations, the Swift-Hohenberg equations, the Reaction-Diffusion equations in gradient forms, the Ginzburg-Landau equations for superconductivity and the porous medium equations.

4. Application to the stability problem

Let $\mathcal{E} : U \subset X \to \mathbb{R}$ be a C^1 functional. Here U is an open subset of the real Banach space X. We consider a critical point $\varphi \in U$ which is a *ground state* in the sense that $\mathcal{E}$ attains its minimum at this point:

$$\mathcal{E}(\varphi) = \inf_{u\in U} \mathcal{E}(u).$$

For $r > 0$ we use U_r to denote the r-neighborhood of φ :

$$U_r := \{u \in X : \|u - \varphi\|_X < r\}.$$

Our first result reads as follows.

THEOREM 4.1. **(Stability Theorem; Theorem 1.2 of Chapter 5)** *Let φ be a ground state of $\mathcal{E}$ with $\mathcal{E}(\varphi) = 0$ and assume the following gradient inequality:*

$$\phi(\mathcal{E}(u)) \le \|\mathcal{E}'(u)\|_{X'} \quad \forall u \in U_\sigma \cap U, \tag{1.27}$$

with some positive constant σ and some ϕ such that $0 \le \frac{1}{\phi} \in L^1_{loc}(\mathbb{R})$.

Let $\mathcal{N} : U \to X$ be a locally Lipschitz map for which there exists a constant $\alpha > 0$ such that

$$\begin{cases} \langle \mathcal{E}'(u), \mathcal{N}(u)\rangle_{X'\times X} \ge \alpha \cdot \|\mathcal{E}'(u)\|_{X'} \cdot \|\mathcal{N}(u)\|_X, \\ \|\mathcal{E}'(u)\|_{X'} \ge \alpha \cdot \|\mathcal{N}(u)\|_X \end{cases} \tag{1.28}$$

for all $u \in U$. Choose $\varepsilon \in (0, \sigma/2)$ to be so small that

$$\int_0^{\mathcal{E}(u)} \frac{ds}{\phi(s)} < (\alpha\sigma)/2 \quad \forall u \in U_\varepsilon. \tag{1.29}$$

Then for each $u_0 \in U_\varepsilon$, the following autonomous Cauchy problem

$$\begin{cases} \dot{u}(t) = -\mathcal{N}(u(t)), t \ge 0; \\ u(0) = u_0 \end{cases} \tag{1.30}$$

admits a global solution $u : [0, \infty) \to U_{\sigma/2}$ that converges as $t \to \infty$ to some stationary point $u_\infty \in \mathcal{N}^{-1}(0) \cap U_\sigma$ with respect to the X norm. As a consequence, φ as an equilibrium of (1.30) is Lyapunov stable.

Moreover, φ is uniformly asymptotically stable if it is an isolated equilibrium and the inequality in the second line of (1.28) is strengthened by

$$(1/\alpha) \cdot \|\mathcal{N}(u)\|_X \ge \|\mathcal{E}'(u)\|_{X'} \ge \alpha \cdot \|\mathcal{N}(u)\|_X \quad (\forall u \in U). \tag{1.31}$$

Our second result is concerned with the convergence in systems generated by pseudo-gradient vector fields associated with the given functional $\mathcal{E}$. Such a system is used to construct solutions to variational problems [**112**] and is a variant of the usual steepest descent method for Hilbert spaces.

We recall [**112**, p. 78] that a C^1 functional $\mathcal{E} : X \to \mathbb{R}$ is said to satisfy the Palais-Smale (in short, (P.-S.)) condition if every sequence $\{u_n\}_{n\geq 1} \subset X$ such that $\sup_{n\geq 1} |\mathcal{E}(u_n)| < \infty$ and $\|\mathcal{E}'(u_n)\|_{X'} \to 0$ as $n \to \infty$ contains a convergent subsequence.

The following Theorem 4.2 (corresponding to Theorem 2.1 of Chapter 5) applies to the following Cauchy problem:

$$\dot{u}(t) = -\mathcal{N}(u(t)) + g(t, u(t)), t \geq 0; \quad u(0) = u_0, \tag{1.32a}$$

where $\mathcal{N} : X \to X$ is a locally Lipschitz map for which there exists a positive constant α such that

$$\begin{cases} \langle \mathcal{E}'(u), \mathcal{N}(u) \rangle_{X' \times X} \geq \alpha \cdot \|\mathcal{E}'(u)\|_{X'}^2, \\ \|\mathcal{E}'(u)\|_{X'} \geq \alpha \cdot \|\mathcal{N}(u)\|_X \end{cases} \qquad \forall u \in X. \tag{1.32b}$$

Moreover, $g : \mathbb{R}_+ \times X \to X$ is a locally Lipschitz map that satisfies the following smallness condition:

$$\begin{cases} \|g(s,u)\|_X \leq g(s) \quad \forall s \geq 0, u \in X; \\ \displaystyle\int_0^\infty \left(\int_t^\infty g(s)^2 \, ds \right)^{1/2} dt < \infty. \end{cases} \tag{1.32c}$$

THEOREM 4.2. **(Theorem 2.1 of Chapter 5)**
Assume that the C^1 functional $\mathcal{E} : X \to \mathbb{R}$ is bounded below, satisfies the (P.-S.) condition, and satisfies a gradient inequality of type $\mathcal{G}_k$ near every point of X.

Then for any $u_0 \in X$, the Cauchy problem (1.32a) admits a global solution $u : [0, \infty) \to X$ that converges as $t \to \infty$ to some critical point of $\mathcal{E}$.

Theorem 4.2 implies particularly the **convergence of the steepest descent method** concerning a functional $\mathcal{E}$ that is bounded below, satisfies the (P.-S.) condition and is such that its gradient satisfies a gradient inequality of type $\mathcal{G}_k$ near any given point of X. More importantly, from the point of view of practical computations, Theorem 4.2 ensures that the afore mentioned convergence of the steepest descent method is stable under perturbations satisfying the smallness condition (1.32c). For the case where the gradient inequalities are of the special LS-type (this is the case if $\mathcal{E}'$ is analytic and Fredholm), then the above smallness condition (1.32c) can be greatly relaxed into the type of (1.20c); i.e., the second line in (1.32c) can be relaxed as

$$\int_0^\infty \rho(t) g(t)^2 \, dt + \int_0^\infty \frac{1}{\rho(t)} < \infty, \tag{1.33}$$

where ρ is some increasing positive function. A smallness condition of the form (1.33) will be satisfied in many practical situations.

5. Additional remarks

(a) In Chapter 2 we will highlight many gradient inequalities for both finite- and infinite-dimensional cases. These gradient inequalities serve to ensure the applicability of the convergence results obtained in the present work.

We will close this chapter by reviewing some other convergence results obtained without using gradient inequalities. Moreover, we will give a short description of a program for further study.

(b) **(Convergence in dissipative systems).** The following convergence result should be known, but we had difficulty in finding suitable sources. A related result for Hilbert spaces without perturbation can be found in the monograph of J. W. Neuberger [**91**, Theorem 4.12, p. 26].

THEOREM 5.1. *Let E be a Banach space such that its dual E' is strictly convex. Let $\mathcal{M} : U \subset E \to E$ be an accretive map, i.e.,*

$$\langle \mathcal{M}(u) - \mathcal{M}(v), \mathcal{F}(u - v)\rangle \geq 0 \quad \forall u, v \in U, \tag{1.34a}$$

where $\mathcal{F} : E \to E'$ is the duality map given by

$$\mathcal{F}(w) := \{f \in E' : \langle w, f\rangle = \|w\|_E^2 = \|f\|_{E'}^2\} \quad (w \in E).$$

Let $u : [0, \infty) \to U$ be a global solution of the following perturbed dissipative system

$$\dot{u}(t) = -\mathcal{M}(u(t)) + g(t, u(t)),\ t \geq 0, \tag{1.34b}$$

where the perturbation $g : \mathbb{R}_+ \times E \to E$ is locally Lipschitz and satisfies the following smallness condition:

$$\sup_{u \in E} \|g(t, u)\|_E \leq g(t) \quad \forall t \geq 0; \quad \int_0^\infty g(t)\, dt < \infty. \tag{1.34c}$$

If a subsequence of $\{u(t)\}_{t \geq 0}$ converges to some equilibrium of (1.34b), i.e., if there exists an unbounded increasing sequence $\{t_n\}_{n \geq 1}$ and some $\varphi \in E$ such that

$$\mathcal{M}(\varphi) = 0 \quad \text{and} \tag{1.35a}$$

$$\|u(t_n) - \varphi\|_E \to 0 \quad \text{as} \quad n \to \infty, \tag{1.35b}$$

then $\|u(t) - \varphi\|_E \to 0$ as $t \to \infty$.

PROOF. The strict convexity of E' implies [**32**, Prop. 13.1, p. 124] that the function $h(t) := \frac{1}{2}\|u(t) - \varphi\|_E^2$ $(t \geq 0)$ is differentiable and

$$h'(t) = \langle \dot{u}(t), \mathcal{F}(u(t) - \varphi)\rangle.$$

Substituting $\dot{u} = -\mathcal{M}(u) + g(t, u)$ and using $\mathcal{M}(\varphi) = 0$, we find that

$$h'(t) = -\langle \mathcal{M}(u(t)) - \mathcal{M}(\varphi), \mathcal{F}(u(t) - \varphi)\rangle + \langle g(t, u), \mathcal{F}(u(t) - \varphi)\rangle.$$

The first sum term is non positive by the accretivity of $\mathcal{M}$, while the second sum term is bounded by $g(t)(2h(t))^{1/2}$. Therefore,

$$h'(t) \leq 2g(t)h(t)^{1/2} \quad \forall t \geq 0. \tag{1.36a}$$

Integrating (1.36a), we find that

$$h(t)^{1/2} \leq h(t_n)^{1/2} + \int_{t_n}^{t} g(s)\, ds \tag{1.36b}$$

for all $t \geq t_n$. Taking the limit $t \to \infty$ in (1.36b), we obtain that

$$\limsup_{t\to\infty} h(t)^{1/2} \leq h(t_n)^{1/2} + \int_{t_n}^{\infty} g(s)\, ds \tag{1.36c}$$

for all $n \in \mathbb{N}$. By (1.34c) and (1.35b) we have $h(t_n) \to 0$ and $\int_{t_n}^{\infty} g(s)\, ds \to 0$ as $n \to \infty$. Therefore, $h(t) \to 0$ as $t \to \infty$ by (1.36c). This is the desired convergence. □

(c) (**Other known convergence results**) For $\Omega \subset \mathbb{R}^N$ with $N \geq 2$, a number of sufficient conditions are known that guarantee the long-time convergence of solutions to the semilinear heat equation (1.1).

Let L_φ be the self-adjoint operator

$$L_\varphi := \Delta - f_u(x, \varphi(x)) : D(\Delta) \subset L^2(\Omega) \to L^2(\Omega),$$

which is obtained by linearizing the right-hand side of Eq. (1.1) about an equilibrium φ of (1.1). Let $\omega(u_0)$ be the ω-limit set of a given solution u of Eq. (1.1). If zero happens to be an eigenvalue of L_φ with multiplicity at most one for each $\varphi \in \omega(u_0)$, then P.-L. Lions [**79**] proves the long-time convergence. Lions' method takes advantage of comparison results using the maximum principle, and his idea is closely related to an abstract result due to J. K. Hale and P. Massatt [**52**], which has been proved by J. K. Hale and G. Raugel [**53**, Theorem 2.4] in a more rigorous manner. Theorem 2.4 in [**53**] applies to the so-called *thin domains* $\Omega \subset \mathbb{R}^2$ for which the property that the multiplicity is at most one holds true and thus the convergence follows; see also [**100**]. A thin domain in $\mathbb{R}^2$ is a sufficiently thin neighborhood of an arc.

When Ω is a ball and f is independent of $x \in \Omega$, A. Haraux and P. Poláčik [**62**] prove the long-time convergence of nonnegative solutions.

E. Feireisl [**36**] also obtains some results concerning the long-time behavior of solutions to Eq. (1.1) in the total space $\Omega = \mathbb{R}^N$.

The semilinear heat equation (1.1) belongs to the so-called class of *monotone dynamical systems* [**109, 125**]. M. Hirsch shows that the generic convergence holds for such systems. The interested reader should consult the monographs of Hal L. Smith [**109**] and X. Q. Zhao [**125**] and the references therein.

(d) (**A program for further studies**) Below we sketch a program for further studies concerning the applications in the following topics: geometric evolution problems, image processing, phase-field models, and optimization problems.

• Applications to the unique tangent cone problem for minimal surfaces and to the unique tangent problem for harmonic maps due to L. Simon [**107, 108**]. What Simon needs for these applications is the convergence result for the following second-order gradient-like and structurally excited systems

$$\ddot{u} - \dot{u} = -\mathcal{E}'(u) + g(t, u, \dot{u}), \quad t \geq 0, \tag{1.37a}$$

where the energy functional $\mathcal{E}$ is given by an analytic function $E : \bar{\Omega} \times \mathbb{R}^{1+n} \to \mathbb{R}$ as follows:

$$\mathcal{E}(u) = \int_{\Omega} E(x, u, \nabla u)\, dx. \tag{1.37b}$$

Moreover, $E(x, u, \xi)$ is assumed to be uniformly convex in the ξ variable so that the linearization of the gradient map $\mathcal{M}(u) := \mathcal{E}'(u)$ about any equilibrium of the system (1.37a) is a second-order linear elliptic operator. This ellipticity is very essential in establishing gradient inequalities for $\mathcal{M}$. Note that systems of the form (1.37a) fit to our systems (1.21) with more general functionals $\mathcal{E}$. Therefore, applications of Theorems 4.4 and 4.5 of Chapter 3 to other geometric problems should be expected.

We point out that M. F. Bidaut-Veron and L. Veron [**120, 121**] have used the ideas of L. Simon to study the conformal asymptotics of the isothermal gas spheres equations, and P. F. Guang and G. F. Wang [**50**] are able to prove their convergence result for a fully nonlinear conformal flow using a variant of Simon's convergence result given by B. Andrews [**10**].

• Application to the "geometric studies of trajectories of gradient fields". The geometric studies for trajectories of *analytic* gradient vector fields have been done by several authors; see e.g., K. Kurdyka, T. Mostowski and A. Parusiński [**76**], A. Nowel and Z. Szafraniec [**92**], F. Santz [**103**], and the references therein. One main tool used (and needed) by these authors is the afore mentioned gradient inequality of Łojasiewicz for analytic gradient maps. It seems that the ideas used by these authors can be extended to *non-analytic* cases where some suitable kinds of gradient inequalities (e.g., those used in our present work) are available.

• Applications to the "calculus of variations" [**112, 122, 43, 27**], in particular, applications to those problems arising from geometric interests [**9, 112, 107, 108**].

• Application to image processing. As proposed by L. Alvarez, F. Guichard, P.-L. Lions and J.-M. Morel [**5**] (the AGLM-theory), the mathematical equations involved in image processing are the following (non-linear) evolution equations:

$$\dot{u} = F(D^2u, Du, u, t). \tag{1.38}$$

In [**5**], the forms for F are not specified and might be very arbitrary. In many practical uses [**12**], a gradient system (1.38) with an energy of the form

$$\mathcal{E}(u) = \int_\Omega |u_0 - Ru|^2\,dx + \lambda \int_\Omega \rho(|\nabla u|)\,dx \tag{1.39}$$

serves to ensure good digital quality. If the function ρ is convex, then our convergence results apply and ensure the convergence of compact solutions. However, as reported in [**12**], a surprising numerical fact is that one can observe very good numerical results using nonconvex functions ρ (e.g., $\rho(s) = s^2/(1+s^2)$) or functions ρ (e.g., $\rho(s) = s^p$ with $p \neq 2$) such that the corresponding gradient systems are degenerate. The failure of convexity of the functions ρ (and thus the failure of ellipticity of linearization of the gradient map about an equilibrium) makes the mathematical treatment of the convergence problem very difficult. Results in this direction are very rare. It seems that the result of E. Feireisl and F. Simondon [**40**] for the porous medium equations is the only such known results. Thus, a further deep study is needed.

• Applications to phase-field models [**15**]. In the very recent works [**2, 3, 4, 37, 38**] of S. Aizicovici, E. Feireisl, H. Petzeltová and F. Issard-Roch, the convergence to equilibria of solutions to many phase-field models with analytic nonlinearities are established, by virtue of certain suitable gradient inequalities. These results provide the prototype for further studies.

• Application to optimization problems. In solving optimization problems, there are two kinds of *steepest descent methods*: the continuous one and the discrete one. As seen in Theorem 4.2, the validity of gradient inequalities ensures the convergence to one single critical point of the continuous steepest descent method. However, from the point of view of practical computation, the problem concerning the convergence of the discrete steepest descent method is more interesting. In the recent work [**68**] the author himself has been able, for the finite-dimensional case, to establish the result that the validity of gradient inequalities implies the convergence to one single limit for minimizing sequences generated by discrete gradient methods. We expect that the ideas used in [**68**] can be extended to the infinite-dimensional case. It might be thought that the convergence result for the continuous gradient methods implies, to some extent, the convergence of discrete gradient methods. However, as shown in [**68**], there are very crucial differences between the continuous and discrete gradient methods.

CHAPTER 2

Gradient inequality

A gradient inequality links the gradient of a given functional and the functional itself by comparing their increments. As explained in the previous chapter, gradient inequalities play a crucial role in proving convergence results for evolutionary equations. It is our aim in this chapter to give a systematic study of this topic. One of the main results in this chapter is Theorem 4.2 from which many known and some new gradient inequalities will be derived.

1. Basic properties of gradient maps

We recall some useful properties of gradient maps.

1A: Differentiable and analytic maps. Let X, Y be two Banach spaces. As usual, we use $L(X, Y)$ to denote the Banach space of all bounded linear operators from X into Y.

Let $F : U \to Y$ be a map, where $U \subset X$ is open. Consider $u \in U$. Recall that F is *Gâteaux differentiable* at u with a Gâteaux derivative $F'(u) \in L(X, Y)$ if

$$\lim_{t\to 0} t^{-1}\|F(u+th) - F(u) - F'(u)th\|_Y = 0 \quad \forall h \in X.$$

Furthermore, if F is Gâteaux differentiable at u and there holds

$$\lim_{h\to 0} \|F(u+h) - F(u) - F'(u)h\|_Y / \|h\|_X = 0,$$

then F is said to be *Fréchet differentiable* at u. It is well known (cf. [**32**, Prop. 7.5, p. 47]) that if F is Gâteaux differentiable near u and the Gâteaux derivative is continuous at u, then F is Fréchet differentiable at u.

Recall that F is called (real) *analytic* at u if there exist a constant $r > 0$ and continuous symmetric n-linear forms $\{A_n\}$ such that $\sum_{n\geq 1} \|A_n\| r^n < \infty$ and the equation $F(u+h) = F(u) + \sum_{n\geq 1} A_n(h^n)$ holds for small h. If F is differentiable/analytic at every point $u \in U$, then F is said to be differentiable/analytic. As shown in [**32**, Prop. 15.2, p. 150], in case where both X and Y are complex Banach spaces, a map $F : U \to Y$ is analytic if and only if F is Fréchet differentiable.

1B: Gradient maps. In the sequel, U denotes an open subset in a real Banach space $(X, \|\cdot\|_X)$. Moreover, Y is a subspace of the dual space X', which becomes a Banach space under its own norm $\|\cdot\|_Y$. Hence, the embedding $(Y, \|\cdot\|_Y) \hookrightarrow (X', \|\cdot\|_{X'})$ is continuous.

DEFINITION 1.1. A continuous map $\mathcal{M} : U \to Y$ is called a *gradient map* if there exists a C^1 functional $\mathcal{E} : U \to \mathbb{R}$ such that $\mathcal{M}(u) = \mathcal{E}'(u)$ for all $u \in U$, i.e.,

$$\mathcal{E}'(u)h = \langle \mathcal{M}(u), h\rangle \quad \forall u \in U, h \in X,$$

where $\langle \cdot, \cdot \rangle$ is the canonical bilinear form on $X' \times X$. Correspondingly, $\mathcal{E}$ is called a *potential* for the map $\mathcal{M}$.

It follows that if $\mathcal{E}_1$ and $\mathcal{E}_2$ are two potentials for a given gradient map $\mathcal{M}$, then the difference $\mathcal{E}_1 - \mathcal{E}_2$ is constant on every connected component of U. In the literature (see, e.g., [**13, 32**]), gradient maps are also called *gradient mappings/operators* or *potential mappings/operators.*

We have the following basic characterization for gradient maps.

PROPOSITION 1.2. *Let $U \subset X$ be open and convex and assume $\mathcal{M} : U \to Y \subset X'$ to be a continuous map. Define the anti-symmetric form $\{u, v\}$ on U by the integral*

$$\{u, v\} := \int_0^1 \langle \mathcal{M}(tu + (1-t)v), u - v\rangle\, dt \quad (u, v \in U).$$

(i) If $\mathcal{M}$ is a gradient map associated with a C^1 potential $\mathcal{E} : U \to \mathbb{R}$, then there holds

$$\mathcal{E}(u) - \mathcal{E}(v) = \{u, v\} \quad \forall u, v \in U.$$

(ii) $\mathcal{M}$ is a gradient map if and only if the following three-point identity holds:

$$\{u, v\} + \{v, w\} + \{w, u\} = 0 \quad \forall u, v, w \in U.$$

(iii) If $\mathcal{M}$ is of class C^1, then $\mathcal{M}$ is a gradient map if and only if all of its Fréchet derivatives $\mathcal{M}'(u)$ $(u \in U)$ are symmetric in the sense that

$$\langle \mathcal{M}'(u)v, w\rangle = \langle \mathcal{M}'(u)w, v\rangle \quad \forall u, v, w \in U.$$

As result, a bounded linear operator $A : X \to X'$ is a gradient operator if and only if A is symmetric. In this case, a potential for A is given by $\mathcal{E}_A(u) := \frac{1}{2}\langle Au, u\rangle$ $(u \in X)$.

(iv) If $\mathcal{M}$ is an analytic gradient map, then any potential $\mathcal{E} : U \to \mathbb{R}$ such that $\mathcal{M} = \mathcal{E}'$ is analytic as well.

The proof of Proposition 1.2 (i) is easy; the proofs for both (ii) and (iii) can be found in [**13**, pp. 94–95]. The proof of (iv) is elementary and hence will be omitted.

2. Gradient inequality

Warning! In the context followed, we will frequently truncate our notation of "$\mathcal{M} : U \to Y$ is a gradient map associated with a potential $\mathcal{E}$" with "$\mathcal{M} = \mathcal{E}' : U \to Y$ is a gradient map". Note her that the meaning of the equality "$\mathcal{M} = \mathcal{E}'$" is not exact–although the gradient map $\mathcal{M}$ and the Fréchet derivative $\mathcal{E}'$ of the C^1 functional $\mathcal{E}$ share the same representation form, their topological properties may be different. For example, the continuity and analyticity for $\mathcal{M}$ are considered under the norm in Y, not in X', while for $\mathcal{E}'$ these properties are considered under the norm in X'.

Our main subject in this section is the new notion *gradient inequality*. A gradient inequality in a given subset for a gradient map is a comparison of increments of the gradient map and its potential over that given subset. As it will become clearer later, gradient inequalities work as essential tools for proving convergence results for gradient-like trajectories, and for different types of trajectories, we need different types of gradient inequalities. These types of gradient inequalities are determined by the following different classes of functions.

DEFINITION 2.1. Let

$$\mathcal{G} := \{\phi : 0 \leq \frac{1}{\phi} \in L^1_{loc}(\mathbb{R})\}.$$

Equivalently, $\mathcal{G}$ is the class of Lebesgue measurable and nonnegative functions ϕ defined on $\mathbb{R}$ such that the reciprocal $\frac{1}{\phi}$ has a finite Lebesgue integral over every finite interval.

(i) Let $k \geq 1$ be a constant. We denote by $\mathcal{G}_k$ the subclass of $\mathcal{G}$ consisting of all $\phi \in \mathcal{G}$ for which there holds the estimate

$$\phi(x+y) \leq k\phi(x) + |y|^{1/2} \quad \forall x, y \in \mathbb{R}.$$

(ii) A function $\phi \in \mathcal{G}$ is called *quasi-monotone* if there exists some constant $c > 0$ such that

$$\phi(y) \geq c \cdot \phi(x) \quad \forall 0 \leq x \leq y \leq 2x.$$

The class $\mathcal{G}$ is closed under finite minimization. More precisely, the minimization

$$\phi(x) := \inf\{\phi_i(x) : 1 \leq i \leq N\} \quad (x \in \mathbb{R})$$

of each finite collection $\{\phi_i : 1 \leq i \leq N\} \subset \mathcal{G}$ belongs again to $\mathcal{G}$. This follows from the inequality $1/\phi(x) \leq \sum_{i=1}^N 1/\phi_i(x)$ ($\forall x \in \mathbb{R}$). Moreover, it is also easily verified that the class $\mathcal{G}_k$ (for each fixed $k \geq 1$) is closed again under finite minimization. Typical examples of quasi-monotone functions are those given by $\phi_\theta(x) := c \cdot |x|^\theta$ ($x \in \mathbb{R}$) with exponents $\theta \in (0,1)$ and constants $c > 0$. At the end of this chapter we will give a more detailed discussion of these functions. In particular, we will see that every quasi-monotone function in $\mathcal{G}$ dominates a member in $\mathcal{G}_1$.

To have some concrete examples, we consider the class of functions $\phi \in \mathcal{G}$ for which there exist constants $c \in (0, 1/2]$, $\theta \in [1/2, 1)$ and $a \in \mathbb{R}$ such that

$$\phi(x) = c \cdot \min\{|x-a|^\theta, |x-a|^{1/2}\} \quad \forall x \in \mathbb{R}.$$

Note that

$$\phi(x+y) \leq c \cdot \min\{(|x-a| + |y|)^\theta, (|x-a| + |y|)^{1/2}\}.$$

Therefore, if $|x-a| \geq |y|$, then

$$\phi(x+y) \leq 2c \cdot \min\{|x-a|^\theta, |x-a|^{1/2}\} = 2\phi(x);$$

if $|x-a| < |y|$, then

$$\phi(x+y) \leq 2c \cdot \min\{|y|^\theta, |y|^{1/2}\} \leq (2c)|y|^{1/2} \leq |y|^{1/2}.$$

Here we have used the assumption that $c \leq 1/2$. In conclusion, we have

$$\phi(x+y) \leq 2\phi(x) + |y|^{1/2} \quad (\forall x, y \in \mathbb{R})$$

and thus $\phi \in \mathcal{G}_2$.

More complicated quasi-monotone functions in $\mathcal{G}$ are those given by $\phi(s) := \text{const} \cdot |s|(-\log|s|)^\alpha$ for all $|s| \leq \gamma < 1$ and $\phi(s)$ is identical to the constant $\phi(\gamma)$ for all $|s| > \gamma$. Here the constant $\alpha > 1$ is such that these functions are in $\mathcal{G}$.

DEFINITION 2.2. Let $\mathcal{M} = \mathcal{E}' : U \to Y$ be a gradient map in the sense of Definition 1.1.

(i) Given a subset V of U, we say that $\mathcal{M}$ satisfies a *gradient inequality in V* (in short, *g.i. in V*) if there exists some $\phi \in \mathcal{G}$ such that the following inequality holds:

$$(GI) \qquad \phi(\mathcal{E}(v)) \leq \|\mathcal{M}(v)\|_Y \quad \forall v \in V.$$

Accordingly, we say that $\mathcal{M}$ satisfies a gradient inequality near a given point in U if it satisfies a gradient inequality in a neighborhood of that point.

The gradient inequality (GI) is said to be of *type* $\mathcal{G}_k$ if the function ϕ involved in (GI) belongs to the subclass $\mathcal{G}_k$.

(ii) The gradient inequality near the given point $\varphi \in U$ is said to be of *LS-type* (standing for Łojasiewicz-Simon type) if it has the form

$$c \cdot |\mathcal{E}(u) - \mathcal{E}(\varphi)|^\theta \leq \|\mathcal{E}'(u)\|_Y \quad \forall u \in U, \|u-\varphi\|_X < \sigma,$$

where $c, \sigma > 0$ and $\theta \in [1/2, 1)$ are constants.

(iii) The gradient inequality near the given point $\varphi \in U$ is said to be of *weak LS-type* if it has the form

$$\phi(|\mathcal{E}(u) - \mathcal{E}(\varphi)|) \leq \|\mathcal{E}'(u)\|_Y \quad \forall u \in U, \|u-\varphi\|_X < \sigma,$$

where $\sigma > 0$ is a constant and $\phi \in \mathcal{G}$ is a quasi-monotone function.

Recall that the set of critical points of $\mathcal{E}$ is defined by

$$\mathcal{S} := \{\varphi \in U : \mathcal{E}'(\varphi) = 0\} = \{\varphi \in U : \mathcal{M}(\varphi) = 0\}.$$

Points in $U \setminus \mathcal{S}$ are called *regular*. An easy argument yields that $\mathcal{M}$ satisfies a g.i. near each regular point. Thus, the study of gradient inequality is a matter concerned with the study of the behavior of $\mathcal{E}$ near its critical points.

In the above definition, the use of the terminology *weak LS-type* is certainly very inaccurate because the class of quasi-monotone functions is clearly far more complicated than the class of functions $c|\cdot - a|^\theta$ involved in the LS-type gradient inequalities. We decided to use it for reasons of simplicity and because of the following property shared commonly by both types of gradient inequalities: If a gradient map satisfies a gradient inequality of weak LS-type near a critical point φ, then there is only one critical value, namely the value $\mathcal{E}(\varphi)$, in some neighborhood of φ; i.e., we must have $\mathcal{E}(\varphi_1) = \mathcal{E}(\varphi)$ for any critical point φ_1 that is sufficiently near φ. This follows from the observation that a quasi-monotone function $\phi \in \mathcal{G}$ is such that $\phi(x) > 0$ for all $x > 0$.

3. Finite-dimensional gradient inequality

THEOREM 3.1. *Let U be an open subset of $\mathbb{R}^N$, $x_0 \in U$ and let $\Gamma : U \to \mathbb{R}$ be a real valued function.*

(i) (**S. Łojasiewicz**, 1963) *If Γ is* **real analytic** *in a neighborhood of x_0, then Γ satisfies a gradient inequality of LS-type near x_0. More precisely, there exist $\theta \in [1/2, 1)$, $\gamma \geq 2$ and $\sigma > 0$ such that*

$$|\nabla\Gamma(x)| \geq |\Gamma(x) - \Gamma(x_0)|^\theta + \operatorname{dist}(x, W)^\gamma \quad \forall x \in \mathbb{R}^N, |x - x_0| < \sigma,$$

where $W := \{x \in \mathbb{R}^N : |x - x_0| < \sigma, \nabla\Gamma(x) = 0\}$.

(ii) (**L. Simon**, 1996) *Assume Γ to be a C^2 function and $\nabla\Gamma(x_0) = 0$. If the connected component C of the critical points set $\{x \in U : \nabla\Gamma(x) = 0\}$ that contains x_0 has the same dimension as the kernel of the Hessian matrix $\mathrm{Hess}_\Gamma(x_0)$ of Γ at x_0 locally near x_0, and x_0 lies in the interior of the component C, then there holds the following gradient inequality of LS-type with the best possible exponent $\theta = 1/2$:*

$$|\nabla\Gamma(x)| \geq c \cdot |\Gamma(x) - \Gamma(x_0)|^{1/2} \quad \forall x \in \mathbb{R}^N, |x - x_0| < \sigma,$$

with two positive constants c and σ.

Here and later, the norm in a Euclidean space is denoted by $|\cdot|$.

Theorem 3.1 (i) is well known and was established by S. Łojasiewicz [**81**] in 1963 and proved in 1965 [**82**, Prop. 1, p. 92] (also cf. [**83**]). Theorem 3.1 (ii) was observed by L. Simon in [**108**, p. 80] and was clarified in the above form recently by R. Chill and M. A. Jendoubi in [**21**, Prop. 5.4 (ii)] (cf. also [**61**]). Simon calls the condition given as above an *integrability condition* [**108**, p. 80] since it is equivalent to the hypothesis that there is a smooth d-dimensional manifold of the critical points set $\{x \in U : \nabla\Gamma(x) = 0\}$, where d is the dimension of the kernel of the Hessian matrix $\mathrm{Hess}_\Gamma(x_0)$, which is tangent to the kernel of $\mathrm{Hess}_\Gamma(x_0)$. Such an integrability condition is certainly difficult to check.

Interested readers would probably like to know the difficulty encountered by Łojasiewicz in proving his inequality. To have a real feel, it is suggested to try to determine the (Łojasiewicz) exponent $\theta \in [1/2, 1)$ for the following real analytic function $V(x, y) := x(x^2 - y^3)$ $(x, y \in \mathbb{R})$ that has the unique critical point at the origin. The task is to find an exponent $\theta \in [1/2, 1)$ (whose existence is ensured by the above Łojasiewicz theorem) such that const$\cdot|V(x,y)|^\theta \le |\nabla V(x,y)|$ for all $(x,y) \in [-1,1]^2$, i.e., to find an exponent $\theta \in [1/2, 1)$ such that

$$\text{const} \cdot |x(x^2 - y^3)|^\theta \le \sqrt{(3x^2 - y^3)^2 + (3xy^2)^2} \quad \forall x, y \in [-1, 1].$$

(Hint: Consider the restriction on the elliptic curve $y^3 = 3x^2$. It yields the first requirement that $\theta \ge 7/9$. It was shown by Prof. G. Metafune (Univ. Lecce) that $\theta = 7/9$ suffices.)

It seems that the above two results cited in Theorem 3.1 are the unique known generic criteria on the validity of gradient inequalities for functions. Therefore, the following one-dimensional result, due to the author himself, is certainly interesting.

THEOREM 3.2. *Let $V \subset \mathbb{R}^1$ be a bounded interval. Then every C^1 function $f : V \to \mathbb{R}$ satisfies a gradient inequality in V.*

To prove Theorem 3.2, we need an extra construction. Let $V \subset \mathbb{R}^1$ be a bounded interval and consider a C^1 function $f : V \to \mathbb{R}$. Recall that a value $z \in \mathbb{R}$ is a *regular value* of f if $f^{-1}(z) \cap \Lambda = \emptyset$, and a *critical value* otherwise, where $\Lambda := \{x : f'(x) = 0\}$ is the set of critical points. It follows from Sard's theorem [**104**] that the set $f(\Lambda)$ of critical values has Lebesgue measure zero. Moreover, using the Inverse Function Theorem, one can easily establish that the preimage $f^{-1}(z)$ of a regular value z contains at most finite points.

We define a nonnegative function ϕ_f on the bounded interval $f(V)$ by

$$\phi_f(z) := \inf\{|f'(x)| : x \in V \cap f^{-1}(z)\}, \quad z \in f(V).$$

The function ϕ_f establishes the following inequality connecting f and its derivative:

$$\phi_f(f(x)) \le |f'(x)| \quad \forall x \in V.$$

Moreover, we note that ϕ_f vanishes identically on $f(V \cap \Lambda)$ and is strictly positive on $f(V) \setminus f(V \cap \Lambda)$. We extend ϕ_f to be a nonnegative function defined on $\mathbb{R}$ by setting $\phi_f(x) = \infty$ for all $x \notin f(V)$. Further properties of this function follow.

LEMMA 3.3. *The function ϕ_f defined as above obeys the following continuity and integrability:*

(i) ϕ_f is continuous on the set $f(V) \setminus f(V \cap \Lambda)$ of regular values. Hence, ϕ_f is lower semicontinuous on $f(V)$.

(ii) The Lebesgue integral

$$I(\phi_f) := \int_{f(V)} \frac{dz}{\phi_f(z)}$$

exists and is bounded by the Lebesgue measure of V, i.e., $I(\phi_f) \le \lambda_1(V)$.

PROOF. The continuity in (i) can be proved in a routine way. Here we omit the details. By (i) the function $1/\phi_f$ is continuous outside the set of critical values $f(V \cap \Lambda)$, which has Lebesgue measure zero. Therefore, $1/\phi_f$ is Lebesgue measurable.

Case I: V is compact. We first establish a weaker estimate:

$$(*) \qquad I(\phi_f) \leq \lambda_1(V) + \lambda_1(f(V)).$$

To prove $(*)$, we recall the following formula that represents the Lebesgue integral of a nonnegative function g defined on $\mathbb{R}$ by a Riemann integral:

$$\int_{\mathbb{R}} g(x)\,dx = \int_0^\infty \lambda_1(\Lambda_t)\,dt,$$

where λ_1 is the Lebesgue measure on $\mathbb{R}$ and the sets Λ_t are defined by

$$\Lambda_t := \{x \in \mathbb{R} : g(x) > t\}.$$

For a proof of this formula, we refer to Maz'ja [**87**, pp. 36–37].

Now we apply the above formula to the function $g = 1/\phi_f$ restricted in the bounded interval $f(V)$ and obtain that

$$I(\phi_f) = \int_0^\infty F(t)\,dt \quad \text{with} \quad F(t) := \lambda_1(\Lambda_t) \quad \text{and}$$

$$\Lambda_t := \{z \in f(V) : \phi_f(z)^{-1} > t\} = \{z \in f(V) : \phi_f(z) < 1/t\}, \quad t \geq 0.$$

Since F is nonincreasing, we have that $I(\phi_f) \leq \sum_{n=0}^\infty F(n)$. For each $k \in \mathbb{N}$, we define a set

$$S_k := \{z \in f(V) : 1/(k+1) \leq \phi_f(z) < 1/k\}.$$

The collection $\{S_k : k \in \mathbb{N}\}$ forms a class of disjoint measurable subsets of $f(V)$. Moreover, for each $n \in \mathbb{N}$, the set Λ_n is decomposed as $\Lambda_n = \bigcup_{k=n}^\infty S_k$. It follows that $F(n) = \sum_{k=n}^\infty \lambda_1(S_k)$ and thus

$$\sum_{n=1}^\infty F(n) = \sum_{k=1}^\infty k\lambda_1(S_k).$$

We claim that

$$S_k \subseteq f(T_k) \quad \text{with} \quad T_k := \{x \in V : 1/(k+1) \leq |f'(x)| < 1/k\}.$$

To see this, let $z_0 \in S_k$. Then we can find a sequence $(x_n)_{n\geq 1}$ in $V \cap f^{-1}(z_0)$ such that

$$\lim_{n\to\infty} |f'(x_n)| = \phi_f(z_0) \in [1/(k+1), 1/k).$$

Since V is compact, we can assume further that $(x_n)_{n\geq 1}$ converges to a point $x_0 \in V$. This implies by the continuity of f and f' that $f(x_0) = \lim_{n\to\infty} f(x_n) = z_0$ and

$$|f'(x_0)| = \lim_{n\to\infty} |f'(x_n)| = \phi_f(x_0) \in [1/(k+1), 1/k).$$

In other words, $x_0 \in T_k$ and thus $z_0 \in f(T_k)$. Hence $S_k \subseteq f(T_k)$ and thus $\lambda_1(S_k) \leq \lambda_1(f(T_k))$. The collection $\{T_k : k \in \mathbb{N}\}$ forms a class of disjoint Baire subsets of V. Moreover, by the famous lemma of Sard [**104**], each image $f(T_k)$ is Lebesgue measurable and

$$\lambda_1(f(T_k)) \leq \int_{T_k} |f'(x)|\,dx.$$

(Note that the assumption that the underlying set V is one-dimensional is crucial for the above estimate, since such an estimate is generally false for the higher dimensional case!) Since $|f'(x)| < 1/k$ for all $x \in T_k$, we have

$$\int_{T_k} |f'(x)|\, dx \leq (1/k) \cdot \int_{T_k} dx = \lambda_1(T_k)/k.$$

It follows that

$$k\lambda_1(S_k) \leq k\lambda_1(f(T_k)) \leq \lambda_1(T_k) \quad \forall k \in \mathbb{N}.$$

Consequently,

$$\sum_{n=1}^{\infty} F(n) = \sum_{k=1}^{\infty} k\lambda_1(S_k) \leq \sum_{k=1}^{\infty} \lambda_1(T_k) = \lambda_1\left(\bigcup_{k=1}^{\infty} T_k\right) \leq \lambda_1(V).$$

Since $F(0) = \lambda_1(f(V))$, we have

$$I(\phi_f) \leq F(0) + \sum_{n=1}^{\infty} F(n) \leq F(0) + \lambda_1(V) = \lambda_1(f(V)) + \lambda_1(V).$$

This proves $(*)$.

Next we derive the desired result $I(\phi_f) \leq \lambda_1(V)$ by a scaling trick. Fix a constant $\alpha > 0$ and consider on the compact interval $U := \frac{1}{\alpha}V$ the C^1 function g given by

$$g(y) := f(\alpha y) \quad (y \in U).$$

Replacing the pair (V, f) by the pair (U, g), we construct the function ϕ_g as above. A calculation yields

$$\phi_g(z) = \alpha\phi_f(z) \quad \forall z \in g(U) = f(V).$$

Therefore,

$$I(\phi_g) = \int_{g(U)} \frac{dz}{\phi_g(z)} = \frac{1}{\alpha} \int_{f(V)} \frac{dz}{\phi_f(z)} = \frac{1}{\alpha} I(\phi_f).$$

On the other hand, since U is compact, we are allowed to apply the estimate in $(*)$. It yields that

$$I(\phi_g) \leq \lambda_1(g(U)) + \lambda_1(U) = \lambda_1(f(V)) + \lambda_1(V)/\alpha.$$

Consequently, we have that

$$I(\phi_f) = \alpha I(\phi_g) \leq \lambda_1(V) + \alpha\lambda_1(f(V))$$

for all $\alpha > 0$. Letting $\alpha \to 0$ we obtain finally that

$$(**) \qquad I(\phi_f) \leq \lambda_1(V).$$

Case II: V is not compact. We choose a sequence $\{V_n : n \in \mathbb{N}\}$ of compact intervals $V_n \subset V$ such that $V_n \subset V_{n+1}$ for all $n \in \mathbb{N}$ and $V = \bigcup_n V_n$. For each V_n we let ϕ_n be the function that is equal to ∞ outside $f(V_n)$ and is given on $f(V_n)$ by

$$\phi_n(z) := \inf\{|f'(x)| : x \in V_n \cap f^{-1}(z)\}, \quad z \in f(V_n).$$

Since each V_n is a compact interval, we have from $(**)$ that

$$\int_{f(V_n)} \frac{1}{\phi_n(z)}\, dz \leq \lambda_1(V_n) \leq \lambda_1(V) < \infty.$$

On the other hand, we have that

$$\lim_{n\to\infty} \frac{1}{\phi_n(z)} = \frac{1}{\phi_f(z)} \geq 0$$

for all $z \in \mathbb{R}$. By Fatou's Lemma, this implies that

$$I(\phi_f) = \int_{f(V)} \frac{dz}{\phi_f(z)} \leq \liminf_{n\to\infty} \int_{f(V_n)} \frac{dz}{\phi_n(z)} \leq \lambda_1(V) < \infty.$$

This completes the proof of Lemma 3.3. □

Now Theorem 3.2 follows directly from Lemma 3.3. In fact, Lemma 3.3 yields that $\phi_f \in \mathcal{G}$ and $\phi_f(f(x)) \leq |f'(x)|$ for all $x \in V$. Hence, f satisfies a gradient inequality in V.

For the one-dimensional case we also have the following result that applies to the class of *quasi-analytic* functions [**74**]. The proof (using the Taylor expansion) is easy and thus will be omitted.

PROPOSITION 3.4. *Let $f : (-1,1) \to \mathbb{R}$ be a C^{k+1} function such that $f'(0) = 0$ and $f^{(j)}(0) = 0$ for all $j \leq k$ but $f^{(k+1)}(0) \neq 0$. Then, near $x = 0$, the derivative f' satisfies a gradient inequality of the following type:*

$$c \cdot |f(x)|^{\theta} \leq |f'(x)| \quad \forall x \in (-1,1), |x| \leq \sigma,$$

where $\theta := k/(k+1)$ and both c and σ are positive constants.

To close this section we also give some two-dimensional gradient inequalities.

EXAMPLE 3.5. We consider a C^1 function $V : \mathbb{R}^2 \to \mathbb{R}$ that is given, in the polar coordinates $(x, y) = (r\cos\theta, r\sin\theta)$, by

$$V(x,y) \equiv V(r,\theta) := A(r)g(B(r) + \theta),$$

where the three functions A, B, and g obey the following properties:

(i) (**Conditions on** A, B) Both A and B are C^1 functions. Let $[\delta, a]$ be a given interval, $0 \leq \delta < a$, for which there exists an increasing function $\phi \in \mathcal{G}$ such that the function $s/\phi(s)$ is nondecreasing for all $s > 0$ and there holds

$$\phi(|A(r)|) \leq |A'(r)| \quad \forall r \in [\delta, a]. \tag{2.0a}$$

(ii) (**Condition on** g) $g : \mathbb{R} \to \mathbb{R}$ is a C^1 function, 2π-periodic and

$$\max\{|g(s)|, |g'(s)|\} \geq 1 \quad \forall s \in \mathbb{R}.$$

Typical examples of functions g satisfying condition (ii) are the appropriate scalar products of sine and cosine functions.

Below we want to show that the conditions (i,ii) ensure that V satisfies a gradient inequality with an increasing function in $\mathcal{G}$ in the ring domain

$\{z \in \mathbb{R}^2 : \delta \le |z| \le a\}$. More precisely, we want to show that there holds the gradient inequality

$$\tilde{\phi}(|V(x,y)|) \le |\nabla V(x,y)| \quad \forall r = \sqrt{x^2+y^2} \in [\delta, a] \tag{2.0b}$$

with some function $\tilde{\phi} \in \mathcal{G}$ that is increasing in $\mathbb{R}_+$. In (2.0b), the interval $[\delta, a]$ is the same as the one involved in (2.0a).

For this purpose, we assume (2.0a). We need to find the function $\tilde{\phi}$ by virtue of the function ϕ. Note that $|\nabla V(x,y)|^2 = |V_r|^2 + |V_\theta/r|^2$ and thus

$$|\nabla V(x,y)| \ge \max\{|V_r|, |V_\theta|/a\} \quad \forall r \le a. \tag{2.0c}$$

We have, with $s := B(r) + \theta$, that

$$V_r = A'(r)g(s) + A(r)B'(r)g'(s), \quad V_\theta := A(r)g'(s).$$

Since B is a C^1 function, there exists some constant $b \ge 1$ such that $|B'(r)| \le b$ for all $r \le a$. It follows that $|V_r| + b|V_\theta| \ge |A'(r)g(s)|$ and thus, by (2.0c), that

$$|\nabla V(x,y)| \ge c_1 \cdot \max\{|A'(r)g(s)|, |A(r)g'(s)|\} \quad \forall r \le a,$$

where $c_1 := \min\{1/2, 1/(2ab)\} > 0$ is independent of the functions A and g. Next we use (2.0a) to find that $|A'(r)| \ge \phi(|A(r)|)$ for all $r \in [\delta, a]$. Consequently, we have that

$$|\nabla V(x,y)| \ge c_1 \cdot \max\{\phi(|A(r)|)|g(s)|, |A(r)g'(s)|\} \tag{2.0d}$$

for all $r \in [\delta, a]; s \in \mathbb{R}$. Comparing (2.0d) with (2.0b), we see that we need to find an increasing function $\tilde{\phi} \in \mathcal{G}$ such that

$$\tilde{\phi}(uv) \le c_1 \cdot \max\{\phi(u)v, u|g'(s)|\} \tag{2.0e}$$

for all $u := |A(r)|, v := |g(s)|$ and $r \in [\delta, a]$, $s \in \mathbb{R}$.

By condition (ii) we have that $\max\{v, |g'(s)|\} \ge 1$.

We assume first that the function $\tilde{\phi}$ has been chosen such that $\tilde{\phi} \le c_1\phi$. Then, if $v := |g(s)| \ge 1$, we have that

$$\frac{\tilde{\phi}(uv)}{uv} \le \frac{c_1\phi(uv)}{uv} \le \frac{c_1\phi(u)}{u},$$

since $uv \ge u$ and the function $0 < s \mapsto s/\phi(s)$ is nondecreasing. From this we find that

$$\tilde{\phi}(uv) \le c_1\phi(u)v \quad \text{if} \quad v \ge 1.$$

As a consequence, we see that (2.0e) holds true for the case $v = |g(s)| \ge 1$, provided that $\tilde{\phi} \le c_1\phi$. Consider the other case $v = |g(s)| < 1$. Then we have that $|g'(s)| \ge 1$. Hence, (2.0e) will be satisfied if there holds

$$\tilde{\phi}(uv) \le c_1 \cdot \max\{\phi(u)v, u\} \quad \forall u \ge 0, v \ge 0. \tag{2.0e$'$}$$

To continue the search for the function $\tilde{\phi}$ satisfying (2.0e$'$), we define a function $\psi : \mathbb{R}_+ \to \mathbb{R}_+$ by setting $\psi(0) := 0$ and

$$\psi(s) := s^2/\phi(s) \quad \forall s > 0. \tag{2.0f}$$

By the monotonicity of the function $s/\phi(s)$ for $s > 0$ combined with the property $\phi \in \mathcal{G}$, we find that ψ is a continuous and strictly increasing function on $\mathbb{R}_+$. Let $\phi_1 := \psi^{-1}$ be the inverse of ψ, i.e., the strictly increasing function $\phi_1 : \mathbb{R}_+ \to \mathbb{R}_+$ is determined by the equation

$$\phi_1(\psi(s)) = s \quad \forall s \geq 0. \tag{2.0g}$$

We have that

$$I := \int_0^{\psi(1)} \frac{d\tau}{\phi_1(\tau)} = \int_0^1 \frac{d\psi(s)}{\phi_1(\psi(s))} = \int_0^1 \frac{d\psi(s)}{s}.$$

Thus, a partial integration yields that

$$I = \psi(1) + \int_0^1 \frac{\psi(s)}{s^2}\, ds = \psi(1) + \int_0^1 \frac{ds}{\phi(s)} < \infty,$$

since $\phi \in \mathcal{G}$ (i.e., $0 \leq \frac{1}{\phi} \in L^1_{loc}(\mathbb{R})$) and $\psi(s)/s = s/\phi(s) \to 0$ as $s \downarrow 0$. Therefore, the increasing function ϕ_1 is such that $0 \leq \frac{1}{\phi_1} \in L^1_{loc}(\mathbb{R}_+)$ and thus, by setting $\phi_1(s) := \phi_1(-s)$ for $s < 0$, we obtain a function $\phi_1 \in \mathcal{G}$.

We show that ϕ_1 satisfies a variant of (2.0e′), namely, that it satisfies the inequality

$$\phi_1(uv) \leq \max\{\phi(u)v, u\} \quad \forall u, v \geq 0. \tag{2.0h}$$

Clearly, (2.0h) holds true if $uv = 0$. Conversely, if (2.0h) does not hold for all $u \geq 0, v \geq 0$, then there exist $u > 0, v > 0$ such that

$$\phi_1(uv) > \phi(u)v, \quad \phi_1(uv) > u.$$

On the one hand, since $\phi_1 = \psi^{-1}$ in $\mathbb{R}_+$, we find from the inequality $\phi_1(uv) > u$ that

$$uv > \psi(u) = u^2/\phi(u),$$

or equivalently,

$$w := v\phi(u) > u.$$

On the other hand, we find from the inequality $\phi_1(uv) > \phi(u)v = w$ that

$$uv > \psi(w) = \frac{w}{\phi(w)} \cdot w = \frac{w}{\phi(w)} \cdot v\phi(u),$$

or equivalently,

$$\frac{u}{\phi(u)} > \frac{w}{\phi(w)}.$$

This is a contradiction, since the monotonicity of the function $s/\phi(s)$ for $s > 0$ combined with the property $w > u$ implies that $u/\phi(u) \leq w/\phi(w)$. Therefore, we have proved (2.0h). It follows that the function

$$\tilde{\phi} := \min\{\phi, \phi_1\} \in \mathcal{G}$$

is increasing on $\mathbb{R}_+$, and satisfies the inequalities (2.0e, e′) and thus the desired inequality (2.0b).

From the above proof we see that there holds

$$\{z \in \mathbb{R}^2 : \nabla V(z) = 0\} \subset \{z \in \mathbb{R}^2 : A(|z|)A'(|z|) = 0\}.$$

REMARK 3.6. (a) The functions $\phi \in \mathcal{G}$ considered in Example 3.5 have the monotonicity that both $\phi(s)$ and $s/\phi(s)$ are increasing for all $s \geq 0$. Typical functions satisfying this monotonicity are those given by $\phi(s) := \text{const}\cdot|s|^\theta$ ($s \in \mathbb{R}$) with $\theta \in (0,1)$, which are used to define gradient inequality of LS-type, and those given by $\phi(s) := c \cdot |s|(-\log|s|)^\alpha$ for $|s| \leq \gamma < 1$ and $\phi(s) := (c\gamma)(-\log\gamma)^\alpha$ for all $|s| > \gamma$ with some $\gamma \in (0,1)$ and $\alpha > 1$.

(b) The function A_1 given by $A_1(r) := 0$ for all $r \in [0,1]$ and $A_1(r) := \exp(-1/(r-1))$ for all $r > 1$ satisfies, in every compact interval of $\mathbb{R}_+$, a gradient inequality of the form (2.0a) with a function $\phi \in \mathcal{G}$ given by $\phi(s) := \text{const} \cdot |s|(\log|s|)^2$ for $|s| \leq 1/e$; $\phi(s)$ is identical to the constant $\phi(1/e)$ for all $|s| > 1/e$. This function has the property that both $\phi(s)$ and $s/\phi(s)$ are increasing for $s \geq 0$. Therefore, the result in Example 3.5 implies that a C^∞ potential V_1 of the form $V_1(r,\theta) = A_1(r)g(B(r)+\theta)$ with a C^∞ function B and a 2π-periodic C^∞ function g such that $\max\{|g(s)|, |g'(s)|\} \geq 1$ satisfies, in every compact subset of $\mathbb{R}^2$, a gradient inequality of weak LS-type. Note that the set of critical points of V_1 is equal to the closed unit disk and the potential V_1 is totally degenerate at these critical points in the sense that all derivatives of V_1 vanish at these critical points.

Functions V of the form $V(r,\theta) = A(r)g(B(r)+\theta)$ have been used by J. Palis and W. de Melo [**93**, pp. 14–15] to construct planar gradient systems that admit nonconvergent bounded trajectories. One important factor in their nonconvergent example is that the function B is *unbounded*; the non-convergent trajectory approaches spirally to the unit circle with infinitely many turns. The unboundedness of the turns is given by the unboundedness of the function $B(r)$, which becomes unbounded as $r \to 1$.

Below we give a more direct method for constructing functions with the afore mentioned nonconvergence property. Our construction reveals that the omega-limit set of the nonconvergent bounded trajectories can be very arbitrary. In fact, for any given connected and closed subset of the unit circle, there always exists a C^∞ function V on $\mathbb{R}^2$ such that the corresponding two-dimensional gradient system $\dot{z} = -\nabla V(z)$ admits a bounded trajectory whose omega-limit set is equal to the given connected and closed subset of the unit circle.

EXAMPLE 3.7. We consider C^2 functions $V : \mathbb{R}^2 \to \mathbb{R}$ that have the following form in the polar coordinates (r,θ):

$$V(r,\theta) := a(r) + b(r)\sin(\theta - w(r)). \tag{2.1a}$$

We assume that $V(r,\theta) \equiv 0$ for all $r \geq 1$. It follows that, for any initial value $z_0 = (x_0, y_0)$ with $|z_0| < 1$, the gradient system

$$\dot{z} = -\nabla V(z) \tag{2.1b}$$

has a unique global solution $z(t)$ such that $z(0) = z_0$ and $|z(t)| < 1$ for all $t \geq 0$.

We look for relations among the three functions a, b, and w that imply that (2.1b) has a special solution of the form $\theta = w(r)$, i.e., a solution $z(t)$

with the following representation in the polar coordinates:

(2.1c) $$z(t) = (r(t)\cos w(r(t)), r(t)\sin w(r(t))).$$

Using the equivalent representation $\dot{r} = -\frac{\partial V}{\partial r}$, $\dot{\theta} = -\frac{1}{r}\frac{\partial V}{\partial \theta}$ of (2.1b) in the polar coordinates $z = (r\cos\theta, r\sin\theta)$, an elementary manipulation (whose details are omitted) yields that (2.1b) admits a special solution of the form (2.1c), provided that the functions a, b, and w satisfy the following relation:

(2.1d) $$b(r) = ra'(r)w'(r)/(1 + rw'(r)^2).$$

Using (2.1d), the equation for the radius function $r(t)$ is

(2.1e) $$\dot{r} = (-a'(r))/(1 + rw'(r)^2).$$

Next, we further assume the following property for the function $a(r)$: There exists some $\tau \in [0, 1)$ such that $a'(r) < 0$ for all $r \in [\tau, 1)$. It follows from this that the following curve

(2.1f) $$\zeta := \{(r\cos w(r), r\sin w(r)) : r \in [\tau, 1)\}$$

is an invariant set for the gradient system (2.1b), and the omega-limit set of every solution $z(t)$ of (2.1b) with an initial value $z(0) \in \zeta$ is equal to the subset $\omega(\zeta)$ of the unit circle, which consists of all points $(\cos\theta, \sin\theta)$ for which there exists a sequence $(r_n)_{n\in\mathbb{N}} \subset [0, 1)$ such that $r_n \to 1$ and $(\cos w(r_n), \sin w(r_n)) \to (\cos\theta, \sin\theta)$ as $n \to \infty$. Since the function w is arbitrary, the invariant curve ζ as well as the omega-limit set $\omega(\zeta)$ can be arbitrarily pre-assigned.

(a) Let Λ be a connected and closed subset of the unit circle. By a result [**65**, Lemma 2.1] (also cf. [**64**, Lemma]) from this author, there exists a C^∞ function $\tilde{w} : [0, 1) \to [0, 2\pi)$ such that the omega-limit set as $r \to 1$ of the map $r \mapsto (r\cos\tilde{w}(r), r\sin\tilde{w}(r))$ is equal to Λ. Moreover, there exists another C^∞ function $\tilde{a} : [0, 1) \to \infty$ such that $\tilde{a}'(r) < 0$ for all $r \in [1/2, 1)$, and

(*) $$\tilde{a}^{(m)}(0) = 0, \quad \lim_{r\to 1^-} \tilde{a}^{(m)}(r)\tilde{w}^{(n)}(r) = 0$$

for all $m, n \in \mathbb{N} \cup \{0\}$. The growth conditions in (*) imply that the function $\tilde{b}(r) := r\tilde{a}(r)\tilde{w}'(r)/(1+r\tilde{w}'(r)^2)$ $(r \in [0, 1))$ given by (2.1d) is a C^∞ function on $[0, 1)$. Finally, the function $\tilde{V}$ of the form (2.1a) with the choice $(a, b, w) := (\tilde{a}, \tilde{b}, \tilde{w})$ is a C^∞ function with the property that the curve $\tilde{\zeta}$, given by (2.1f) under the choice $\tau := 1/2$ and $w := \tilde{w}$, is an invariant set for the gradient system $\dot{z} = -\nabla\tilde{V}(z)$ and every solution of it with an initial value in $\tilde{\zeta}$ takes the given connected and closed subset Λ as its omega-limit set.

(b) For the special case where Λ is the total unit circle, more transparent functions a, b, w can be chosen as follows. We take $w(r) := 1/(1 - r^2)$ and $a(r) := \exp(-1/(1 - r^2))$. Then the function b determined by (2.1d) is

$$b(r) := -4r^3 e^{-\frac{1}{1-r^2}}/(4r^3 + (1 - r^2)^4)$$

for all $r \in [0,1)$, and the corresponding V function given by (2.1a), denoted by V_2, has the form

$$V_2(r,\theta) = e^{-\frac{1}{1-r^2}} \cdot \Big[1 - \frac{4r^3}{4r^3 + (1-r^2)^4} \cdot \sin\big(\theta - \frac{1}{1-r^2}\big)\Big]$$

for all $r \in [0,1)$ and $\theta \in \mathbb{R}$. The shape of the function V_2 can be thought of as a "Mexican hat". The corresponding invariant curve

$$\zeta = \{(r\cos(1-r^2)^{-1}, r\sin(1-r^2)^{-1}) : r \in [0,1)\}$$

takes the total unit circle as its omega-limit set as $r \to 1$, and thus every solution $z(t)$ of the gradient system $\dot{z} = -\nabla V_2(z)$ with an initial value $z(0) \in \zeta$ takes the total unit circle as its omega-limit set.

We point out the following connection to gradient inequalities. A function V constructed above does not satisfy a gradient inequality (of any type in our sense) near every critical point located in the unit circle, since, as we have already seen in Chapter 1 (Theorem 1.2), the validity of a gradient inequality near some cluster point of a bounded (and thus precompact) trajectory of the corresponding gradient system $\dot{z} = -\nabla V(z)$ will force the trajectory to converge as $t \to \infty$ to a single critical point. The above function V_2 and a function V_1 considered in Remark 3.6 (b) have the following in common: Both are totally degenerate at the critical points located in the unit circle. However, V_1 does satisfy a gradient inequality in every compact subset, while V_2 does not. This reveals that it is generally a hard task to establish gradient inequalities, except for the one-dimensional case (Theorem 3.2) and the *analytic case* (Theorem 3.1 (i), Łojasiewicz's Theorem).

4. Infinite-dimensional gradient inequality

Before describing the next construction result, we make a preparation. Recall that a bounded linear operator $L : X_1 \to X_2$ between two Banach spaces X_1 and X_2 is called a *Fredholm operator of index zero* if L has a closed range $R(L)$, a finite dimensional kernel $N(L)$, and

$$\dim N(L) = \dim (X_2/R(L)) < \infty.$$

Recall that

$$R(L) := \{Lu : u \in X_1\} \quad \text{and} \quad N(L) := \{u \in X_1 : Lu = 0\}.$$

Similarly, a C^1 map $\mathcal{M} : U \subset X_1 \to X_2$ is called a *Fredholm map of index zero* if all of its Fréchet derivatives are Fredholm operators of index zero.

We put the following hypotheses **(H1)**–**(H3)** on the quintuplet

$$(H, A, \mathcal{M}, X, \tilde{X}).$$

(H1) H is a Hilbert space and

$$A : D(A) \subset H \to H$$

is a linear self-adjoint and positive definite operator. Set

$$H_A := (D(A), \langle \cdot, \cdot \rangle_A)$$

to be the Hilbert space endowed with the scalar product

$$\langle u, v\rangle_A := \langle Au, Av\rangle_{H\times H} \quad (u, v \in D(A)).$$

(H2) Both X and $\tilde{X}$ are Banach spaces under their own norms and such that the embeddings

$$X \hookrightarrow H_A, \quad \tilde{X} \hookrightarrow H$$

are continuous. Moreover, $X \subset \tilde{X}$.

(H3) $\mathcal{M} = \mathcal{E}' : U \subset X \to \tilde{X}$ is a C^l (resp., analytic) gradient map with $l \geq 1$, having the following properties (1)–(2):

(1) $\mathcal{M}$ is a Fredholm map of index zero; i.e., for each fixed $u \in U$, the bounded linear operator

$$\mathcal{M}'(u) : X \to \tilde{X}$$

is a Fredholm operator of index zero.

(2) For each fixed $u \in U$, the bounded linear symmetric operator

$$\mathcal{M}'(u) : X \to \tilde{X}$$

has an extension

$$\mathcal{M}_1(u) : H_A \to H,$$

which is also a symmetric Fredholm operator of index zero from H_A into H, and such that the map

$$U \ni u \mapsto \mathcal{M}_1(u) \in L(H_A, H) \quad \text{is continuous,}$$

or equivalently, the map $U \ni u \mapsto \mathcal{M}_1(u)A^{-1} \in L(H)$ is continuous.

Here and later, we use the notation $L(X_1, X_2)$ to denote the Banach space of all bounded linear operators from the Banach space X_1 to the Banach space X_2. Moreover, if $X_2 = X_1$, then $L(X_1)$ will be used instead.

Under hypotheses **(H1)**–**(H3)** we consider a critical point φ of $\mathcal{E}$, $\varphi \in U$. Without loss of generality we may assume $\varphi = 0$ and thus $\mathcal{M}(0) = 0$.

Consider

$$G := \mathcal{M}_1(\varphi) = \mathcal{M}_1(0) \tag{2.2a}$$

which is, by **(H3)** (2), a symmetric Fredholm operator of index zero from H_A into H. Therefore, we have the following decomposition:

$$H = N(G) \oplus R(G), \quad N(G) \perp R(G). \tag{2.2b}$$

Let P be the orthogonal projection of H onto $N(G)$. It is easily verified that the sum $P + G : H_A \to H$ is injective as well as surjective and thus is a bijection from H_A onto H. Hence, both operators

$$T := (P + \mathcal{M}_1(0))^{-1}, \quad Q := AT \tag{2.2c}$$

are bounded operators on H. This boundedness of T and Q will be used later.

Next we consider the bounded linear operator

$$L := \mathcal{M}'(0) : X \to \tilde{X}. \tag{2.3a}$$

L is, by **(H3)** (1), a Fredholm operator of index zero. Since $G = \mathcal{M}_1(0)$ is an extension of L, we have that

$$N(L) = N(G) \cap X \tag{2.3b}$$

and thus by (2.2b) there holds

$$\begin{cases} X = N(L) \oplus (X \cap R(G)), \\ \tilde{X} = (N(G) \cap \tilde{X}) \oplus R(L), \end{cases} \tag{2.3c}$$

and

$$\dim (N(G) \cap \tilde{X}) = \dim N(L) < \infty. \tag{2.3d}$$

By the inclusion $X \subset \tilde{X}$ (see **(H2)**) we have that

$$N(L) = N(G) \cap X \subset N(G) \cap \tilde{X},$$

which implies by (2.3d) that

$$N(G) \cap \tilde{X} = N(L). \tag{2.3e}$$

Taking together (2.3c) and (2.3e), we obtain the decompositions

$$X = N(L) \oplus (X \cap R(G)), \quad \tilde{X} = N(L) \oplus R(L). \tag{2.4a}$$

Moreover, note that all of the above decompositions are orthogonal in H. In particular, we have that

$$\langle u, v \rangle_{H \times H} = 0 \quad \forall u \in N(L), v \in R(L) \cup (X \cap R(G)). \tag{2.4b}$$

Let $\Pi : X \to \tilde{X}$ be the finite dimensional operator defined to be zero on $X \cap R(G)$ and such that

$$\Pi : N(L) \to N(L) \quad \text{is the identity.} \tag{2.4c}$$

Consider the sum $\Pi + L : X \to \tilde{X}$. We claim that $\Pi + L$ is a bijection. Indeed, this operator has range $N(L) + R(L) = \tilde{X}$. Moreover, if $(\Pi + L)(x + y) = 0$ with $x \in N(L), y \in X \cap R(G)$, then $x = -Ly \in R(L) \cap N(L) = \{0\}$ by (2.4b). This implies $x = 0$ as well as $Ly = 0$. Hence, $y \in (X \cap R(G)) \cap N(L) = \{0\}$ by (2.4b) again, i.e., $y = 0$. Thus, the sum $\Pi + L$ is also injective and hence a bijection.

We consider a new map $\mathcal{N} : U \to \tilde{X}$ given by

$$\mathcal{N}(u) := \Pi u + \mathcal{M}(u), \quad u \in U. \tag{2.5a}$$

$\mathcal{N}$ is also a C^l (resp., analytic) map and $\mathcal{N}'(0) = \Pi + L$ is a bijective operator as shown above. Therefore, the Inverse Theorem (both for the C^l case [**32**, Theorem 15.2, p. 149] and the analytic case [**32**, Theorem 15.3, p. 151]) applies to $\mathcal{N}$, yields a convex neighborhood $B_1 \subset U$ of $\varphi = 0$, a neighborhood $B_2 \subset \tilde{X}$ of $\mathcal{M}'(\varphi) = 0$, and a C^l (resp., analytic) map $\Psi : B_2 \to B_1$ such that

$$\begin{cases} \Psi(\mathcal{N}(u)) = u & \forall u \in B_1, \\ \mathcal{N}(\Psi(f)) = f & \forall f \in B_2. \end{cases} \tag{2.5b}$$

The convexity of B_1 ensures the validity of the integral representations

$$\mathcal{E}^{(m+1)}(u) - \mathcal{E}^{(m+1)}(v) = \int_0^1 \mathcal{M}^{(m)}(su + (1-s)v)(u-v)\, ds$$

$(u, v \in B_1)$ for $m = 1, ..., l$. Below we will freely use these representations.

We recall from Definition 2.2 that the gradient map $\mathcal{E}' : X \to H$ is said to satisfy a gradient inequality of type $\mathcal{G}_k$ near the critical point $\varphi = 0$ if there holds

$$\phi(\mathcal{E}(u)) \le \|\mathcal{E}'(u)\|_H \quad \forall u \in U, \|u - \varphi\|_X < \sigma,$$

where $\sigma > 0$ is a constant and the function ϕ is such that $0 \le \frac{1}{\phi} \in L^1_{loc}(\mathbb{R})$ and such that

$$\phi(x+y) \le k\phi(x) + |y|^{1/2} \quad \forall x, y \in \mathbb{R}$$

with a constant $k \ge 1$.

PROPOSITION 4.1. **(Lyapunov-Schmidt reduction).** *Under hypotheses* **(H1)–(H3)** *we let $\varphi = 0 \in U$ be a critical point of $\mathcal{E}$ (i.e., $\mathcal{M}(\varphi) = 0$) for which the spaces X and $\tilde{X}$ are decomposed as in (2.4a) and the C^l (resp., analytic) homeomorphism Ψ has been constructed as above.*

Let $\{e_1, ..., e_m\}$ ($m = dim\, N(\mathcal{M}'(0)) < \infty$) be a basis of $N(\mathcal{M}'(0))$. Put

$$V := \{\mathbf{x} = (x_1, ..., x_m) \in \mathbb{R}^m : \sum_{j=1}^m x_j e_j \in B_2\} \tag{2.6a}$$

and define a C^l (resp., analytic) functional $\Gamma : V \to \mathbb{R}$ by

$$\Gamma(\mathbf{x}) := \mathcal{E}(\Psi w) \quad \text{with} \quad w := \sum_{j=1}^m x_j e_j \in B_2. \tag{2.6b}$$

If $\nabla\Gamma$ satisfies a gradient inequality of type $\mathcal{G}_k$, i.e., if there hold the estimates

$$\phi(\Gamma(\mathbf{x})) \le |\nabla\Gamma(\mathbf{x})| \quad \forall \mathbf{x} \in V, |\mathbf{x}| < \sigma \tag{2.6c}$$

with some constant $\sigma > 0$ and $\phi \in \mathcal{G}_k$, then $\mathcal{M}$ satisfies a gradient inequality of type $\mathcal{G}_k$ near the critical point $\varphi = 0$ with the function $c_1 \cdot \phi$, where c_1 is some positive constant. More precisely, there holds

$$c_1 \cdot \phi(\mathcal{E}(u)) \le \|\mathcal{M}(u)\|_H \quad \forall u \in U, \|u - \varphi\|_X < \sigma_1 \tag{2.6d}$$

with some positive constants c_1, σ_1.

PROOF. The proof is divided into several steps.

Step I. We first establish the following

Claim: There are smaller neighborhoods $B_1' \subset B_1$ and $B_2' \subset B_2$, where B_1' is convex, and a positive constant C_2 such that

$$\|\mathcal{M}(u) - \mathcal{M}(v)\|_H \le C_2 \cdot \|A(u-v)\|_H \quad \forall u, v \in B_1', \tag{2.7a}$$

$$\|A(\Psi(f) - \Psi(g))\|_H \le C_2 \cdot \|f - g\|_H \quad \forall f, g \in B_2'. \tag{2.7b}$$

To treat (2.7a), we note that

$$\begin{aligned}\mathcal{M}(u)-\mathcal{M}(v) &= \int_0^1 \mathcal{M}'(su+(1-s)v)(u-v)\,ds \\ &= \int_0^1 (\mathcal{M}'(su+(1-s)v)A^{-1})(Au-Av)\,ds.\end{aligned}$$

Thus, in replacing $\mathcal{M}'(su+(1-s)v) \in L(X,\tilde{X})$ by its extension $\mathcal{M}_1(su+(1-s)v) \in L(H_A,H)$, we find that

$$\|\mathcal{M}(u)-\mathcal{M}(v)\|_H \le \|A(u-v)\|_H \cdot \int_0^1 \|\mathcal{M}_1(su+(1-s)v)A^{-1}\|_{L(H)}\,ds.$$

Using the continuity assumption in **(H3)** (2), we infer from the above that there exist a convex neighborhood B_1' and a constant C_2 such that (2.7a) holds.

To treat (2.7b), we consider the map $\mathcal{L}: U \to L(H_A,H)$, given by

$$\mathcal{L}(u) := P + \mathcal{M}_1(u) \quad (u \in U).$$

We have

$$\mathcal{L}(u)_{|X} = \mathcal{N}'(u) = \Pi + \mathcal{M}'(u) \quad \forall u \in U.$$

By differentiating the equality $\mathcal{N}(\Psi(w)) = w$ ($\forall w \in B_2$), we obtain

$$\mathcal{L}(\Psi(w))\Psi'(w) = \mathrm{Id}_{\tilde{X}}.$$

It yields that

$$\Psi'(w) = \mathcal{L}(h)^{-1} \circ \mathrm{Id}_{\tilde{X}} \quad \text{with} \quad h := \Psi(w) \in B_1,$$

provided $\mathcal{L}(h)$ is invertible.

Now we use (2.2c) to see that $\mathcal{L}(0) = P + \mathcal{M}_1(0)$ has an inverse T, which is a bounded operator on H and such that the product $AT = Q$ is again a bounded operator on H. We want to show that this property for $\mathcal{L}(0)$ remains valid for $\mathcal{L}(h)$ if the norm of h is sufficiently small. To this end, we rewrite

$$\mathcal{L}(h) = [1 - \varepsilon(h)Q]\mathcal{L}(0)$$

with $Q = AT$ and

$$\varepsilon(h) := [\mathcal{L}(0) - \mathcal{L}(h)]A^{-1} = \mathcal{M}_1(0)A^{-1} - \mathcal{M}_1(h)A^{-1}.$$

Now the continuity assumption **(H3)** (2) implies that

$$\|\varepsilon(h)\|_{L(H)} \to 0 \quad \text{as} \quad \|h\|_X \to 0.$$

Therefore, we can find some $\delta_0 > 0$ such that

$$\|\varepsilon(h)\|_{H\to H} < 1/(1+2\|Q\|_{L(H)}) \quad \forall h \in X, \|h\|_X < \delta_0. \tag{2.8a}$$

It follows for all $h \in X$ with $\|h\|_X < \delta_0$ that

$$\mathcal{L}(h)^{-1} = T\sum_{k=0}^{\infty}(\varepsilon(h)Q)^k$$

and thus

$$A\Psi'(w) = A\mathcal{L}(h)^{-1} = Q\sum_{k=0}^{\infty}(\varepsilon(h)Q)^k.$$

Therefore, with $C_3 := \|Q\|_{L(H)}$, we obtain

$$\text{(2.8b)} \qquad \|A\Psi'(w)\|_{L(H)} < 2C_3 \quad \forall w = \Psi^{-1}(h) \quad \text{with} \quad \|h\|_X < \delta_0.$$

The set $\{w \in B_2 : \|\Psi(w)\|_X < \delta_0\}$ is an open neighborhood of $0 = \mathcal{M}(\varphi)$; hence it contains a convex and open neighborhood of 0, say B_2'. Consider now $f, g \in B_2'$. Then we have $sf + (1-s)g \in B_2'$ for all $s \in [0,1]$ and thus $\|\Psi(sf + (1-s)g)\|_X < \delta_0$ for all $s \in [0,1]$. Therefore, the estimate in (2.8b) applies to each $w := sf + (1-s)g$ with $s \in [0,1]$ and yields $\|A\Psi'(sf + (1-s)g)\|_{L(H)} \leq 2C_3$ for all $s \in [0,1]$. It follows that

$$\begin{aligned}
\|A(\Psi(f) - \Psi(g))\|_H &\leq \int_0^1 \|A\Psi'(sf + (1-s)g)(f-g)\|_H \, ds \\
&\leq \|f - g\|_H \cdot \int_0^1 \|A\Psi'(sf + (1-s)g)\|_{L(H)} \, ds \\
&\leq (2C_3) \cdot \|f - g\|_H.
\end{aligned}$$

Therefore, (2.7b) holds true for any $C_2 \geq 2C_3$ and thus the above **Claim** has been established.

Step II. Let B_1', B_2' be neighborhoods given by the above **Claim** and put

$$W := B_1' \cap (\Psi\Pi)^{-1}(B_1') \cap \Psi(B_2') \cap \Pi^{-1}(B_2') \ni 0.$$

W is a neighborhood of $\varphi = 0$. We claim that there is a positive constant C_4 such that

$$\text{(2.9a)} \qquad \|\mathcal{M}(\Psi(\Pi u))\|_H \leq C_4 \cdot \|\mathcal{M}(u)\|_H \quad \forall u \in W,$$

$$\text{(2.9b)} \qquad |\mathcal{E}(u) - \mathcal{E}(\Psi(\Pi u))| \leq C_4 \cdot \|\mathcal{M}(u)\|_H^2 \quad \forall u \in W.$$

To see this, we let $u \in W \subset B_1'$ and put $v := \Psi(\Pi u) \in B_1'$. We have

$$\begin{aligned}
\|\mathcal{M}(v)\|_H &\leq \|\mathcal{M}(v) - \mathcal{M}(u)\|_H + \|\mathcal{M}(u)\|_H \\
&\leq C_2 \cdot \|A(v-u)\|_H + \|\mathcal{M}(u)\|_H \quad \text{(by (2.7a))}.
\end{aligned}$$

Moreover, since $\Pi u \in B_2'$ and $\mathcal{N}(u) \in B_2'$, we have

$$\begin{aligned}
\|A(v-u)\|_H &= \|A(\Psi(\Pi u) - \Psi(\mathcal{N}(u)))\|_H \\
&\leq C_2 \cdot \|\Pi u - \mathcal{N}(u)\|_H \\
&= C_2 \cdot \|\mathcal{M}(u)\|_H \quad \text{(by (2.7b))}
\end{aligned}$$

and thus

$$\text{(2.10)} \qquad \|A(v-u)\|_H \leq C_2 \cdot \|\mathcal{M}(u)\|_H.$$

This implies that (2.9a) holds true for any constant $C_4 \geq 1 + C_2^2$.

To show (2.9b), we note that

$$|\mathcal{E}(u)-\mathcal{E}(v)| = \left|\int_0^1 \langle A^{-1}\mathcal{M}(u+t(v-u)), A(v-u)\rangle_{H\times H}\,dt\right|$$
$$\leq \|A^{-1}\|_{L(H)}\cdot\|A(v-u)\|_H\cdot\int_0^1 \|\mathcal{M}(u+t(v-u))\|_H\,dt.$$

Using (2.7a) again we have that

$$\|\mathcal{M}(u+t(v-u))\|_H \leq \|\mathcal{M}(u)\|_H + C_2\cdot\|A(v-u)\|_H$$

for all $t\in[0,1]$. Thus, by using (2.10) we obtain that

$$|\mathcal{E}(u)-\mathcal{E}(v)| \leq \|A^{-1}\|_{L(H)}C_2(1+C_2^2)\cdot\|\mathcal{M}(u)\|_H^2.$$

Therefore, (2.9b) holds if $C_4 \geq \|A^{-1}\|_{L(H)}C_2(1+C_2^2)$ and thus, both (2.9a) and (2.9b) hold true for any constant $C_4 \geq (1+C_2^2)(1+\|A^{-1}\|_{L(H)}C_2)$.

Step III. Now we assume the validity of the gradient inequality (2.6c). Note that

$$\frac{\partial\Gamma(x)}{\partial x_j} = \langle\mathcal{M}(\Psi w), \Psi'(w)e_j\rangle_{H\times H}$$

with $w := \sum_{j=1}^m x_j e_j \in B_2$. We have

$$\left|\frac{\partial\Gamma(x)}{\partial x_j}\right| \leq \|\Psi'(w)e_j\|_H\cdot\|\mathcal{M}(\Psi w)\|_H$$
$$\leq \text{const}\cdot\|\Psi'(w)e_j\|_X\cdot\|\mathcal{M}(\Psi w)\|_H.$$

For the last estimate, we have used the fact that the injection $X\hookrightarrow H$ is continuous. Furthermore, since $\Psi: B_2\to X$ is of C^l class with $l\geq 1$, we can finally find a positive constant C_5 such that

$$|\nabla\Gamma(x)| \leq C_5\cdot\|\mathcal{M}(\Psi w)\|_H$$

for all $\mathbf{x}\in V$ with $w := \sum_{j=1}^m x_j e_j \in B_2$. Thus, by choosing some small $\sigma_2>0$, we obtain from the gradient inequality (2.6c) that

$$\phi(\mathcal{E}(\Psi w)) \leq C_5\cdot\|\mathcal{M}(\Psi w)\|_H \quad \forall w\in N(L)\cap B_2, \|w\|_{\tilde{X}}<\sigma_2. \tag{2.11}$$

Here, $L=\mathcal{M}'(0)$. Consider the neighborhood

$$W_1 := \{u\in W : \|\Pi u\|_{\tilde{X}} < \sigma_2\} \ni 0 = \varphi.$$

Let $u\in W_1$. Then we can apply (2.11) with $w=\Pi u$. It yields by (2.9a) that

$$\phi(\mathcal{E}(\Psi\Pi u)) \leq (C_5C_4)\cdot\|\mathcal{M}(u)\|_H. \tag{2.12}$$

Recall that $\phi\in\mathcal{G}_k$ is such that

$$\phi(x+y) \leq k\phi(x) + |y|^{1/2} \quad \forall x,y\in\mathbb{R}.$$

Therefore, by applying this to the choice $x=\mathcal{E}(\Psi\Pi u)$ and $y=\mathcal{E}(u)-\mathcal{E}(\Psi\Pi u)$ and using (2.12), we find that

$$\phi(\mathcal{E}(u)) \leq (kC_5C_4)\cdot\|\mathcal{M}(u)\|_H + |\mathcal{E}(u)-\mathcal{E}(\Psi\Pi u)|^{1/2}.$$

We use (2.9b) to estimate the last term. It yields

$$\phi(\mathcal{E}(u)) \leq (kC_5C_4 + \sqrt{C_4}) \cdot \|\mathcal{M}(u)\|_H.$$

In conclusion, we have shown that $\mathcal{M}$ satisfies a gradient inequality in the neighborhood W_1 with the function $\phi/(kC_4C_5 + \sqrt{C_4}) \in \mathcal{G}_k$. This completes the proof. □

As seen in Theorem 3.1, analytic functions in finite-dimensional spaces satisfy a gradient inequality with functions ϕ of the form $\phi(s) = c \cdot |s-a|^\theta$ with constants $c > 0$ and $\theta \in [1/2, 1)$. Analogously, we say that the gradient map $\mathcal{M}$ considered above satisfies a *gradient inequality of LS-type* near φ if there exist positive constants $c, \sigma \in (0, 1/2]$ and $\theta \in [1/2, 1)$ such that

$$c \cdot |\mathcal{E}(u) - \mathcal{E}(\varphi)|^\theta \leq \|\mathcal{M}(u)\|_H \quad \forall u \in U, \|u - \varphi\|_X < \sigma. \tag{2.13}$$

Here, the term "LS-type" means "Łojasiewicz-Simon type", because L. Simon gave the extension of Łojasiewicz's result (Theorem 3.1 (i)) to infinite-dimensional spaces concerning analytic semilinear elliptic operators for the first time in 1983 [**107**]. Simon's result has been improved by Rybka and Hoffmann [**101, 102**], Feireisl and Simondon [**40**] and Takáč [**115**] by replacing the L^2 norm used by Simon with the dual Sobolev norms. Jendoubi [**72**] gives a simplified approach to Simon's method and proves gradient inequalities for gradient maps of the semilinear form $\mathcal{M}(u) = -Au + f(x, u)$. Our following abstract result covers these results and can be also applied to more general settings.

THEOREM 4.2. **(Analytic case)**. *Under hypotheses* **(H1)**–**(H3)** *we assume furthermore that the gradient map* $\mathcal{M} = \mathcal{E}' : U \subset X \to \tilde{X}$ *is analytic. Then we have the following assertions:*

(i) $\mathcal{M}$ *satisfies a gradient inequality of LS-type near every critical point* $\varphi \in U$ *of* $\mathcal{E}$; *i.e., for each* $\varphi \in U$ *with* $\mathcal{E}'(\varphi) = 0$ *there exist positive constants* c, σ, *and* $\theta \in [1/2, 1)$ *such that*

$$c \cdot |\mathcal{E}(u) - \mathcal{E}(\varphi)|^\theta \leq \|\mathcal{E}'(u)\|_H \quad \forall u \in U, \|u - \varphi\|_X < \sigma.$$

(ii) $\mathcal{M}$ *satisfies a gradient inequality of type* $\mathcal{G}_2$ *in every compact subset of* U. *More precisely, for any compact subset* $K \subset U$, *there exists some* $\phi \in \mathcal{G}_2$ *such that*

$$\phi(\mathcal{E}(u)) \leq \|\mathcal{E}'(u)\|_H \quad \forall u \in K.$$

PROOF. We use Proposition 4.1 to prove assertion (i). We also consider the case $\varphi = 0$ and $\mathcal{M}(\varphi) = 0$. Note that the analyticity assumption of $\mathcal{M}$ implies that the functional $\mathcal{E}$ and the map Ψ are analytic as well. Therefore, the function Γ defined in (2.6a,b) is analytic. Applying Theorem 3.1 (i) of Łojasiewicz to this analytic function Γ, we conclude that $\nabla\Gamma$ satisfies a gradient inequality of LS-type. The desired result then follows from Proposition 4.1.

To prove assertion (ii), we take a compact $K \subset U$. By assertion (i), for each $u \in K$ there exist a neighborhood $K(u) \subset U$ of u and two constants $c(u) \in (0, 1/2]$, $\theta(u) \in [1/2, 1)$ such that

$$c(u) \cdot |\mathcal{E}(u) - \mathcal{E}(v)|^{\theta(u)} \leq \|\mathcal{E}'(v)\|_H \quad \forall v \in K(u).$$

Now $\{K(u) : u \in K\}$ forms an open cover of K. Hence, the compactness of K yields a finite cover $\{K(u_i)\}_{i=1}^{i=N}$. For each i, $1 \leq i \leq N$, set

$$\phi_i(x) := c(u_i) \cdot \min\{|x - \mathcal{E}(u_i)|^{\theta(u_i)}, |x - \mathcal{E}(u_i)|^{1/2}\} \quad (x \in \mathbb{R}).$$

As seen in the text following Definition 2.1, each ϕ_i belongs to the class $\mathcal{G}_2$. It follows that

$$\phi(x) := \min\{\phi_i(x) : 1 \leq i \leq N\} \quad (x \in \mathbb{R})$$

is of class $\mathcal{G}_2$. Moreover, if $u \in K(u_i)$ for some $1 \leq i \leq N$, then we have

$$\phi(\mathcal{E}(u)) \leq \phi_i(\mathcal{E}(u)) \leq c(u_i) \cdot |\mathcal{E}(u) - \mathcal{E}(u_i)|^{\theta(u_i)} \leq \|\mathcal{E}'(u)\|_H.$$

Consequently,

$$\phi(\mathcal{E}(u)) \leq \|\mathcal{E}'(u)\|_H \quad \forall u \in K,$$

as desired. □

THEOREM 4.3. **(Non-analytic case).** *Under hypotheses* **(H1)–(H3)** *we assume $\varphi \in U$ to be a critical point of $\mathcal{E}$ for which there exists an $n \geq 2$ such that*

(2.14a) $$\mathcal{M}^{(j)}(\varphi) = 0 \quad \forall 2 \leq j < n \quad \text{and}$$

(2.14b) $$\mathcal{M}^{(n)}(\varphi)e^{n+1} \neq 0 \quad \forall e \in N(\mathcal{M}'(\varphi)), e \neq 0.$$

Then $\mathcal{M}$ satisfies a gradient inequality of LS-type near φ with an exponent $\theta = n/(n+1)$.

REMARK 4.4. (a) If $n = 2$, then condition (2.14a) is void. Thus, we only need condition (2.14b) with $n = 2$ which says that $\mathcal{M}''(\varphi)e^3 \neq 0$ for all $e \in N(\mathcal{M}'(\varphi))$, $e \neq 0$. Similar conditions and related results can be found in [**18**, Cor. 4.4].

(b) If $\dim N(\mathcal{M}'(\varphi)) \geq 2$, then the unit ball $E := \{e \in N(\mathcal{M}'(\varphi)) : \|e\|_X = 1\}$ is connected and symmetric (i.e., $-E = E$). Thus, for such a case, the index n satisfying (2.14b) must be odd, say $n = 2k+1$ with $k \in \mathbb{N}$, and the corresponding condition (2.14b) reads as

(2.14b′) $$\begin{cases} \text{either} \quad \mathcal{M}^{(2k+1)}(\varphi)e^{2k+2} > 0 \quad \forall e \in E \\ \quad \text{or} \quad \mathcal{M}^{(2k+1)}(\varphi)e^{2k+2} < 0 \quad \forall e \in E. \end{cases}$$

(c) Consider the special case where $\varphi = 0$ and $\mathcal{M}(u) = -\Delta u + f(x, u)$ such that $f(x, 0) \equiv 0$. Then $\mathcal{M}'(0) = -\Delta + f_u(\cdot, 0)$ is a Schrödinger operator with the potential $V(x) := f_u(x, 0)$ [**106**] and each $\mathcal{M}^{(n)}(0) = f_n(x, 0)\bullet$ for $n \geq 2$ is a multiplication operator, where $f_n(x, u)$ is the nth order partial derivative of f with respect to the u variable. Now our conditions (2.14a,b) for this special case read as follows: There exists some $k \geq 1$ such that

$$f_j(x,0) \equiv 0 \quad \forall 2 \leq j < 2k+1$$

but

$$\int_\Omega f_{2k+1}(x,0)h(x)^{2k+2}\,dx \neq 0 \tag{2.14b$''$}$$

for all non-zero solutions h of the equation $-\Delta h + f_u(x,0)h = 0$; cf. Chill [**18**, Cor. 4.3]. Clearly, (2.14b$''$) will be true under the condition that either $f_{2k+1}(\cdot,0) > 0$ a.e. or $f_{2k+1}(\cdot,0) < 0$ a.e. (A condition of this type says little other than the requirement that the polynomial expansion of the function $u \mapsto f(x,u)$ around $u = 0$ has the property that all of the second-order coefficients to some odd order vanish and the next nearest non-zero coefficient of even order is strictly positive/negative.)

However, the global positivity/negativity condition above can be weakened a lot. Note that a classical deep result of Carleman and Müller [**106**, Theorem C.9.1] (see also the references given there concerning this topic) claims that the Schrödinger operator $-\Delta + f_u(x,0)$ has the so-called *unique continuation property* (u.c.p.) if the potential $f_u(x,0)$ is bounded. (The u.c.p. says that every distributional solution h of the equation $-\Delta h + f_u(x,0)h = 0$ in an open connected set Ω that vanishes near some $x_0 \in \Omega$ is identically zero.) Hence, if we assume further that $f_u(x,0)$ is bounded, then each non-zero distributional solution h of the equation $-\Delta h + f_u(x,0)h = 0$ is such that $h(x)^2 > 0$ a.e. Therefore, in this case, in order for (2.14b$''$) to be true, it suffices to assume either $f_{2k+1}(\cdot,0) \geq 0$ a.e. and $f_{2k+1}(\cdot,0) > 0$ in some subset of positive measure or $f_{2k+1}(\cdot,0) \leq 0$ a.e. and $f_{2k+1}(\cdot,0) < 0$ in some subset of positive measure.

Proof of Theorem 4.3. We want to use Proposition 4.1 to show that $\mathcal{M}$ satisfies a gradient inequality of LS-type with the exponent $\theta = n/(n+1)$. As seen there, it suffices to show that the function Γ defined by (2.6a,b) has this property.

To this end, we may also assume that $\varphi = 0 = \mathcal{M}(\varphi)$. Set $L := \mathcal{M}'(0)$ and put $E := \{e \in N(L) : \|e\|_X = 1\}$. Since $N(L)$ is finite-dimensional, E is a compact set. Fix $e \in E$ and set

$$\gamma(s) := \Psi(se) \quad (se \in B_2).$$

We have $\Pi\gamma(s) + \mathcal{M}(\gamma(s)) = se$ and thus

$$\mathcal{M}(\gamma(s)) = se - \Pi\gamma(s) \in N(L). \tag{2.15a}$$

Taking derivatives on both sides of (2.15a) and setting $s = 0$, we obtain that

$$L\gamma'(0) = e - \Pi\gamma'(0) \in N(L) \cap R(L) = \{0\}.$$

This implies that $\gamma'(0) = e$ and thus

$$\gamma(s) = se + \sum_{2 \leq j < n} s^j h_j + O(s^n). \tag{2.15b}$$

We want to show that $h_2 = h_3 = \cdots = h_n = 0$. To this end, we use the nth order Taylor expansion of $\mathcal{M}$:

$$\mathcal{M}(u) = Lu + \sum_{2\le j<n} \frac{1}{j!}\mathcal{M}^{(j)}(0)u^j + O(u^n).$$

Combined with assumption (2.14a) (i.e., $\mathcal{M}^{(j)}(0) = 0$ for all $2 \le j < n$), yields that

$$\mathcal{M}(u) = Lu + O(u^n). \tag{2.15c}$$

Substituting $u = \gamma(s)$ into (2.15c) and using (2.15a) again, we obtain that

$$\sum_{2\le j<n} s^j(\Pi + L)h_j = O(s^n)$$

and thus $(\Pi + L)h_j = 0$ for all $2 \le j < n$. But $\Pi + L$ is invertible, so we must have $h_j = 0$ for all $2 \le j < n$ and hence

$$\Psi(se) = \gamma(s) = se + O(s^n). \tag{2.16a}$$

Consider the function

$$G(s) := \mathcal{E}(\Psi(se)) = \mathcal{E}(\gamma(s)) \quad (se \in B_2).$$

We use condition (2.14a) again in the following $(n+1)$th order Taylor expansion

$$\mathcal{E}(u) - \mathcal{E}(0) = \frac{1}{2!}\langle Lu, u\rangle + \sum_{j=3}^{n+1} \frac{1}{j!}\mathcal{M}^{(j-1)}(0)u^j + O(\|u\|_X^{n+2})$$

and obtain

$$\mathcal{E}(u) - \mathcal{E}(0) = \frac{1}{2!}\langle Lu, u\rangle + \frac{1}{(n+1)!}\mathcal{M}^{(n)}(0)u^{n+1} + O(\|u\|_X^{n+2}). \tag{2.16b}$$

We substitute $u = \gamma(s) = se + O(s^n)$ (see (2.16a)) into (2.16b) and collect the terms. By observing that $\langle L\gamma(s), \gamma(s)\rangle = O(s^{2n})$ (since $Le = 0$), we finally find that

$$\begin{cases} G(s) = \frac{s^{n+1}}{(n+1)!}\mathcal{M}^{(n)}(0)e^{n+1} + s^{n+2}R_1(s,e), \\ G'(s) = \frac{s^n}{n!}\mathcal{M}^{(n)}(0)e^{n+1} + s^{n+1}R_2(s,e), \end{cases} \tag{2.17a}$$

where R_1, R_2 are two functions that are uniformly bounded for all $e \in E$ and all small s. Since E is compact, condition (2.14b) implies that

$$\inf\{|\mathcal{M}^{(n)}(0)e^{n+1}| : e \in E\} > 0. \tag{2.17b}$$

Using (2.17a) combined with (2.17b), we find that there exist two positive constants c, σ such that

$$c \cdot |G(s)|^{\frac{n}{n+1}} \le |G'(s)| \quad \forall |s| < \sigma, \forall e \in E. \tag{2.18}$$

Using (2.18) combined with the equality $G'(s) = \langle \mathcal{M}(\gamma(s)), \gamma'(s)\rangle$, we find that the function Γ given by (2.6b) satisfies a gradient inequality of LS-type with the exponent $\theta = n/(n+1)$. This completes the proof. □

In the next section we will apply Theorem 4.2 to derive classical gradient inequalities involved in the classical "calculus of variations". More results concerning the case where the operator $-A$ is a strongly elliptic operator on a domain or some power of such operators will be given in Chapter 4.

To close this section, we cite another result that has been proved by Huang and Takáč [**70**, Prop. 3.3] in a method very similar to those used above. Related generalizations can be found in the recent work of Chill [**18**].

THEOREM 4.5. *Let X be a Banach space for which there exists a Hilbert space H such that the embeddings $X \hookrightarrow H \hookrightarrow X'$ are continuous. Let $\mathcal{E}: X \to \mathbb{R}$ be an analytic potential and let φ be a critical point, i.e., $\mathcal{E}'(\varphi) = 0$. Assume that $\mathcal{E}''(\varphi) : X \to X'$ is a (linear) Fredholm operator of index zero. Then there exist positive constants c, σ, and $\theta \in [1/2, 1)$ such that*

$$c \cdot |\mathcal{E}(u) - \mathcal{E}(\varphi)|^\theta \leq \|\mathcal{E}'(u)\|_{X'} \quad \forall u \in X, \|u - \varphi\|_X < \sigma.$$

5. Variational gradient inequality

Let $\Omega \subset \mathbb{R}^n$ be a bounded domain with a Lipschitz boundary $\partial\Omega$ (see [**45**, p. 94] for this term). Denote $D_i := \partial/\partial x_i$ $(1 \leq i \leq n)$. For $u : \Omega \subset \mathbb{R}^n \to \mathbb{R}^m$, we define $Du : \Omega \to \mathbb{R}^{n\times m}$ by

$$Du := (D_1u_1, \cdots, D_nu_1; \cdots; D_1u_m, \cdots, D_nu_m).$$

Let $F : \Omega \times \mathbb{R}^m \times \mathbb{R}^{mn} \to \mathbb{R}$ be a Carathéodory function (i.e., $F(x, u, \xi)$ is measurable in $x \in \bar{\Omega}$ and continuous in $(u, p) \in \mathbb{R}^m \times \mathbb{R}^{mn}$) satisfying the following properties (f1)–(f2):

(f1) (**Uniform analyticity**) F is uniformly analytic in the sense that there exists a neighborhood $V \subset \mathbb{C}^m \times \mathbb{C}^{mn}$ of the real plane $\mathbb{R}^m \times \mathbb{R}^{mn}$ and an extension of F, denoted by $\tilde{F}$, such that $\tilde{F} : \Omega \times V \to \mathbb{C}$ is *locally bounded* in the sense that $\tilde{F}$ is uniformly bounded in the product $\Omega \times V_0$ for every compact subset $V_0 \subset V$, and $\tilde{F}(x, u, \xi)$ is analytic in both u and ξ variables.

(f2) (**Uniform ellipticity condition**) For each $R > 0$, there exists a constant $c \equiv c(R) > 0$ such that for all $(u, \xi) \in \mathbb{R}^m \times \mathbb{R}^{mn}$ with $|u| + |\xi| < R$, there holds

$$F''_{\xi\xi}(x, u, \xi)(v, \eta)^2 \geq c \cdot |v|^2 \cdot |\eta|^2 \quad \forall (x, v, \eta) \in \Omega \times \mathbb{R}^m \times \mathbb{R}^{nm}.$$

Here $F''_{\xi\xi}$ is the Hessian matrix of the function F with respect to the ξ variable.

We remark that (f1) implies that all (Fréchet) derivatives of the analytic extension $\tilde{F}$ are also locally bounded in $\Omega \times V$.

We will choose one of the following typical boundary conditions:

(B1) (Dirichlet boundary condition)

$$u = 0 \quad \text{on} \quad \partial\Omega.$$

(B2) (Neumann boundary condition)

$$\frac{\partial u}{\partial \mathbf{n}} = 0 \quad \text{on} \quad \partial\Omega.$$

(B3) (Periodic boundary condition)

$$\Omega := [0, l_1] \times \cdots \times [0, l_n] \subset \mathbb{R}^n \text{ and } u \text{ is } \Omega\text{-periodic.}$$

Let p be such that

$$p \geq 2, \quad p > n \tag{2.19a}$$

and set

$$X := \{u \in W^{2,p}(\Omega; \mathbb{R}^m) \quad \text{plus some boundary condition}\}. \tag{2.19b}$$

It follows from the Sobolev Embedding Theorem that

$$X \subset C^1(\bar{\Omega}; \mathbb{R}^m), \tag{2.19c}$$

and thus the functional

$$\mathcal{E}_F(u) := \int_\Omega F(x, u, Du)\, dx \quad (u \in X) \tag{2.19d}$$

is well-defined.

Let

$$\mathcal{H} := \{u \in W^{1,2}(\Omega; \mathbb{R}^m) \quad \text{plus some boundary condition}\} \tag{2.20a}$$

which is a closed subspace of the Hilbert space $W^{1,2}(\Omega; \mathbb{R}^m)$ and thus is itself a Hilbert space. It follows that the dual

$$H := \mathcal{H}' \subset W^{-1,2}(\Omega; \mathbb{R}^m) \tag{2.20b}$$

is also a Hilbert space, and such that $\mathcal{H} \subset H$. We take

$$A := 1 - \Delta \quad \text{with} \quad D(A) := \mathcal{H} \tag{2.20c}$$

so that $A = 1 - \Delta : D(A) \subset H \to H$ is a self-adjoint and positive definite operator. Clearly, we have that $H_A = D(A) = \mathcal{H}$.

Choose

$$\tilde{X} := L^p(\Omega; \mathbb{R}^m) \supset X. \tag{2.20d}$$

LEMMA 5.1. *Under the above conditions we have the following assertions:*

(i) For each $u \in X$, we have

$$\mathcal{E}_F'(u) = F_u(\cdot, u, Du) - div\, F_\xi(\cdot, u, Du). \tag{2.21a}$$

(ii) We have $\mathcal{E}_F'(X) \subset \tilde{X}$ and the gradient map $\mathcal{M}_F := \mathcal{E}_F' : X \to \tilde{X}$ is analytic.

(iii) For each $u \in X$, the derivative $\mathcal{M}_F'(u)$ is the strongly elliptic operator with bounded coefficients given by

$$\mathcal{M}_F'(u)v = F_{11}v + F_{12}Dv - div(F_{12}v) - div(F_{22}Dv) \quad \forall v \in X, \tag{2.21b}$$

where

$$\begin{cases} F_{11} := F''_{uu}(\cdot, u, Du), & F_{22} := F''_{\xi\xi}(\cdot, u, Du), \\ F_{12} := F''_{u\xi}(\cdot, u, Du) = F''_{\xi u}(\cdot, u, Du), \end{cases} \tag{2.21c}$$

and $F''_{uu}, F''_{u\xi} = F''_{\xi u}$ *and* $F''_{\xi\xi}$ *are second-order partial derivatives of* $F(x, u, \xi)$ *with respect to the* u *and* ξ *variables.*

(iv) The map

$$X \ni u \mapsto \mathcal{M}'_F(u) \in L(\mathcal{H}, H) \tag{2.21d}$$

is continuous.

PROOF. Representations (2.21a) and (2.21b) follow from direct computations under the use of the boundary conditions. Moreover, it is a standard result that the strongly elliptic operator $\mathcal{M}'(u)$ (with $u \in X$) extends to a bounded linear operator from $\mathcal{H}$ to H and the map $X \ni u \to \mathcal{M}'(u) \in L(\mathcal{H}, H)$ is continuous. Here we omit the details.

It remains to show the analyticity of $\mathcal{M}_F$. To this end, we fix a $u \in X$. Let $\delta > 0$ be a constant, and V_0 be a compact subset of V such that

$$\|v\|_{X_\mathbb{C}} < \delta \Longrightarrow (u(x) + v(x), Du(x) + Dv(x)) \in V_0 \quad \forall x \in \bar{\Omega}. \tag{2.22a}$$

Here $X_\mathbb{C}$ is the complexification of the real Banach space X. The existence of the positive constant δ and the compact subset V_0 is ensured by the fact $X_\mathbb{C} \subset C^1(\bar{\Omega}; \mathbb{C}^m)$; cf. (2.19c).

Fix $v \in X$ with $\|v\|_X < \delta$. Then the following

$$h(v; z) := \tilde{F}_u(\cdot, u + zv, Du + zDv) - \operatorname{div} \tilde{F}_\xi(\cdot, u + zv, Du + zDv) \tag{2.22b}$$

($z \in \mathbb{C}$ with $|z| \leq 1$) defines a map from the unit disk to the Banach space $\tilde{X}_\mathbb{C}$, the complexification of the real Banach space $\tilde{X}$. A straightforward computation shows that h is analytic in the open unit disk with a derivative given by $h'_z(v; z) = F_{11} \cdot + F_{12} D \cdot - \operatorname{div}(F_{12} \cdot + F_{22} D \cdot)$, where F_{ij} has the same form as (2.21c) but with the replacement of (u, Du) by $(u + zv, Du + zDv)$. From this we claim that $\mathcal{M}_F$ is infinitely differentiable at the point u and

$$\mathcal{M}_F^{(k)}(u)v^k = h_z^{(k)}(v; 0) \quad (k \in \mathbb{N})$$

for all $v \in X$ with $\|v\|_X < \delta$.

Note that (2.22a) combined with the local boundedness of the analytic extension $\tilde{F}$ (see (f1)) yields the uniform boundedness

$$M := \sup\{\|h(v; z)\|_{\tilde{X}_\mathbb{C}} : |z| \leq 1, \|v\|_X < \delta\} < \infty.$$

By the Cauchy formula for analytic functions, this implies

$$\|\mathcal{M}_F^{(k)}(u)v^k\|_{\tilde{X}} = \|h_z^{(k)}(v; 0)\|_{\tilde{X}} \leq M \cdot k!$$

for all $k \in \mathbb{N}$ and all $v \in X$ with $\|v\|_X < \delta$. Hence, the analyticity of $\mathcal{M}_F$ at the point u follows. □

By Lemma 5.1 we infer that the quintuplet $(H, A, \mathcal{M}_F, X, \tilde{X})$ given as above satisfies hypotheses **(H1)**–**(H3)** in §4. Moreover, $\mathcal{M}_F$ is analytic. Hence, as a direct consequence of Theorem 4.2, we have the following result that extends a result of L. Simon [**107**] for the scalar case $m = 1$ under the use of the L^2 norm.

THEOREM 5.2. *Under the above circumstances we let $\varphi \in X$ be a critical point of $\mathcal{E}_F$, i.e., $\varphi \in X$ and $\mathcal{E}_F'(\varphi) = 0$. Then there exist positive constants c, σ, and $\theta \in [1/2, 1)$ such that*

$$c \cdot |\mathcal{E}_F(u) - \mathcal{E}_F(\varphi)|^\theta \le \|\mathcal{E}_F'(u)\|_{W^{-1,2}} \quad \forall u \in X, \|u - \varphi\|_X < \sigma.$$

Moreover, $\mathcal{E}_F'$ satisfies a gradient inequality of type $\mathcal{G}_2$ in every compact subset of X. More precisely, for any compact subset $K \subset X$, there exists some $\phi \in \mathcal{G}_2$ such that

$$\phi(\mathcal{E}_F(u)) \le \|\mathcal{E}_F'(u)\|_{W^{-1,2}} \quad \forall u \in K.$$

REMARK 5.3. (a) The results in Theorem 5.2 remain true if we consider, as L. Simon did in [**107**], a compact C^∞ Riemannian manifold Ω.

(b) We point out that there are essential difficulties in extending the results in Theorem 5.2 (for bounded domains) to the case where the domains Ω are unbounded. One of the reasons is the failure of the Fredholm property of the linearizations around critical points, which are linear elliptic operators, in case that the underlying domain Ω is unbounded and has an infinite Lebesgue measure (e.g., $\Omega = \mathbb{R}^n$).

6. Gradient inequality for monotone gradient maps

Let $\mathcal{M} = \mathcal{E}' : U \subset X \to Y \subset X'$ be a gradient map. We assume that $\mathcal{M}$ is *monotone* in the sense that

$$\text{(2.23a)} \quad \langle u - v, \mathcal{M}(u) - \mathcal{M}(v)\rangle_{X \times X'} \ge \|u - v\|_X \cdot \alpha(\|u - v\|_X^2) \quad \forall u, v \in U,$$

where α is a nonnegative and Lebesgue measurable function defined on $\mathbb{R}_+$.

THEOREM 6.1. *Let $\mathcal{M} = \mathcal{E}' : U \subset X \to Y \subset X'$ be a C^1 gradient map. Assume $\mathcal{M}$ to be monotone in the sense that the inequality (2.23a) holds true for some function $\alpha \in \mathcal{G}$ (i.e., $0 \le \frac{1}{\alpha} \in L^1_{loc}(\mathbb{R})$). Then $\mathcal{M}$ satisfies a gradient inequality near every critical point of $\mathcal{E}$ in U. More precisely, we choose for each $\varphi \in U, \mathcal{E}'(\varphi) = 0$, a constant $\sigma > 0$ so small that $U_\sigma := \{u \in X : \|u - \varphi\|_X < \sigma\} \subset U$ and $c := \sup_{u \in U_\sigma} \|\mathcal{M}'(u)\|_{X \to Y} < \infty$. Then there holds the gradient inequality*

$$\text{(2.23b)} \qquad \phi(\mathcal{E}(u)) \le \|\mathcal{M}(u)\|_Y \quad \forall u \in U_\sigma$$

with the function $\phi \in \mathcal{G}$ given by

$$\text{(2.23c)} \qquad \phi(s) := (1/\delta) \cdot \alpha(|s - \mathcal{E}(\varphi)|/(\delta c)), \quad s \in \mathbb{R}.$$

Here δ is the norm of the continuous embedding $(Y, \|\cdot\|_Y) \hookrightarrow (X', \|\cdot\|_{X'})$.

PROOF. Let $v \in X$ be such that $\|v\|_X < \sigma$. Then we have $v + \varphi \in U_\sigma \subset U$ and thus by (2.23a) we have

$$\begin{aligned} \|v\|_X \cdot \alpha(\|v\|_X^2) &\le \|v\|_X \cdot \|\mathcal{M}(v + \varphi) - \mathcal{M}(\varphi)\|_{X'} \\ &\le \delta \cdot \|v\|_X \cdot \|\mathcal{M}(v + \varphi)\|_Y. \end{aligned}$$

Here we have used the assumption $\mathcal{M}(\varphi) = 0$. Hence

$$\text{(2.24a)} \qquad \alpha(\|v\|_X^2) \le \delta \cdot \|\mathcal{M}(v + \varphi)\|_Y.$$

On the other hand, we have the following integral representation:

$$\mathcal{E}(v+\varphi)-\mathcal{E}(\varphi)=\int_0^1(1-t)\langle\mathcal{M}'(tv+\varphi)v,v\rangle_{X'\times X}\,dt.$$

Note that

$$\begin{aligned}|\langle\mathcal{M}'(tv+\varphi)v,v\rangle_{X'\times X}| &\leq \|\mathcal{M}'(tv+\varphi)v\|_{X'}\cdot\|v\|_X\\ &\leq \delta\cdot\|\mathcal{M}'(tv+\varphi)v\|_Y\cdot\|v\|_X\\ &\leq (\delta c)\|v\|_X^2.\end{aligned}$$

Therefore,

$$|\mathcal{E}(v+\varphi)-\mathcal{E}(\varphi)|\leq(\delta c)\|v\|_X^2. \tag{2.24b}$$

Combining (2.24a) and (2.24b), we obtain the estimate $\phi(\mathcal{E}(v+\varphi))\leq\|\mathcal{M}(v+\varphi)\|_Y$ with the function $\phi\in\mathcal{G}$ given in (2.23c). □

7. Optimal gradient inequality in Hilbert spaces

We recall that a gradient map $\mathcal{M}=\mathcal{E}':U\subset X\to Y\subset X'$ is said to satisfy a gradient inequality of LS-type near a critical point φ if the following inequality

$$c\cdot|\mathcal{E}(u)-\mathcal{E}(\varphi)|^\theta\leq\|\mathcal{M}(u)\|_Y\quad\forall u\in U,\|u-\varphi\|_X<\sigma$$

holds true for some constants $c,\sigma\in(0,1/2]$ and $\theta\in[1/2,1)$. It remains unclear as to why the exponent θ should be restricted in the interval $[1/2,1)$. We observe that if the above gradient inequality holds true for some $\theta\in(0,1)$, then it remains true for each $\theta'\geq\theta$ with some appropriate new constants c',σ'. In other words, if a gradient inequality of LS-type holds true for some exponent θ, then it remains true for each exponent $\theta'\geq\theta$. Thus, the smaller the exponent, the better the property of the gradient map.

The following result says, at least in the setting of Hilbert spaces, that the exponent $\theta=1/2$ is the lowest bound for all of the possible exponents. Therefore, a gradient inequality with exponent $\theta=1/2$ should be treated as an optimal one.

PROPOSITION 7.1. *Let $\mathcal{H}$ be a real Hilbert space and let $\mathcal{E}:U\subset\mathcal{H}\to\mathcal{H}$ be a C^2 functional. Let $\varphi\in U$ be a critical point of $\mathcal{E}$ and assume that the self-adjoint operator $\mathcal{E}''(\varphi)$ is injective.*

Then the following assertions are equivalent:

(i) There exist positive constants c,σ such that

$$c\cdot|\mathcal{E}(u)-\mathcal{E}(\varphi)|^{1/2}\leq\|\mathcal{E}'(u)\|_{\mathcal{H}}\quad\forall u\in U,\|u-\varphi\|_{\mathcal{H}}<\sigma. \tag{$*$}$$

(ii) The self-adjoint operator $\mathcal{E}''(\varphi)$ is invertible.

PROOF. For the sake of convenience, the norm on $\mathcal{H}$ is simply denoted by $\|\cdot\|$ below. Using $\mathcal{E}'(\varphi)=0$, we have

$$\mathcal{E}'(u)=\mathcal{E}'(u)-\mathcal{E}'(\varphi)=\int_0^1\mathcal{E}''(tu+(1-t)\varphi)(u-\varphi)\,dt\quad\text{and}$$

$$\mathcal{E}(u)-\mathcal{E}(\varphi)=\int_0^1 (1-t)\langle \mathcal{E}''(tu+(1-t)\varphi)(u-\varphi),(u-\varphi)\rangle\,dt.$$

(i)$\Longrightarrow$(ii): Assume $(*)$. Fix $\varepsilon\in(0,\sigma)$. By applying $(*)$ with $u-\varphi=\varepsilon v$, where $\|v\|\le 1$, and using the above representation formulae, we obtain

$$c\cdot\left|\int_0^1 (1-t)\langle\mathcal{E}''(\varphi+\varepsilon tv)\varepsilon v,\varepsilon v\rangle\,dt\right|^{1/2}\le\left\|\int_0^1 \mathcal{E}''(\varphi+\varepsilon tv)\varepsilon v\,dt\right\|.$$

Killing ε on both sides of the above inequality and then letting $\varepsilon\to 0$, we finally find that

$$(c/\sqrt{2})\cdot|\langle\mathcal{E}''(\varphi)v,v\rangle|^{1/2}\le\|\mathcal{E}''(\varphi)v\|\quad\forall v\in\mathcal{H},\|v\|\le 1. \tag{2.25a}$$

Consider the self-adjoint bounded linear operator $L:=\mathcal{E}''(\varphi)$. By the classical Spectral Theorem, we can realize L as a multiplication operator on a real Hilbert space $L^2_\mu(\Omega)$ as follows:

$$L=V^*M_gV,\quad\text{where}\quad M_gf:=gf\quad\forall f\in L^2_\mu(\Omega),$$

$V:\mathcal{H}\to L^2_\mu(\Omega)$ is a unitary isomorphism, and $g\in L^\infty_\mu(\Omega)$ is a real function. Throughout this realization, inequality (2.25a) reads for M_g as follows:

$$(c^2/2)\cdot\left|\int_\Omega g(\omega)f(\omega)^2\,d\mu\right|\le\int_\Omega g(\omega)^2f(\omega)^2\,d\mu\quad\forall f\in L^2_\mu(\Omega). \tag{2.25b}$$

By assumption, $\ker(M_g)=\ker(L)=\{0\}$. Thus, Ω is decomposed into two parts:

$$\Omega=\Omega_+\cup\Omega_-,\quad\Omega_\pm:=\{\omega\in\Omega:\pm g(\omega)>0\}.$$

By considering $\Omega_\pm$, we find from (2.25b) that

$$\int_\Omega(\pm g(\omega))(c^2/2\mp g(\omega))f(\omega)^2\,d\mu\le 0\quad\forall f\in L^2_\mu(\Omega_\pm).$$

An easy argument yields that $|g(\omega)|\ge c^2/2$ for almost every $\omega\in\Omega$ and thus

$$\|M_gf\|\ge(c^2/2)\cdot\|f\|\quad\forall f\in L^2_\mu(\Omega).$$

This implies the invertibility of M_g and hence the invertibility of L.

(ii)$\Longrightarrow$(i): Assume $L:=\mathcal{E}''(\varphi)$ to be invertible in the sense that

$$\|Lv\|\ge(2\gamma)\cdot\|v\|\quad\forall v\in\mathcal{H}.$$

Choose $\sigma>0$ to be so small that

$$\|\mathcal{E}''(\varphi+v)-L\|\le\gamma\quad\forall v\in\mathcal{H},\|v\|<\sigma.$$

Then, on the one hand we have for all $v\in\mathcal{H}$, $\|v\|<\sigma$, that

$$\|\mathcal{E}'(v+\varphi)-Lv\|=\left\|\int_0^1(\mathcal{E}''(\varphi+tv)-L)v\,dt\right\|\le\gamma\|v\|.$$

Consequently,

$$\|\mathcal{E}'(v+\varphi)\|\ge\|Lv\|-\gamma\|v\|\ge\gamma\|v\|\quad\forall v\in\mathcal{H},\|v\|<\sigma. \tag{2.25c}$$

On the other hand, we have for all $v \in \mathcal{H}$, $\|v\| < \sigma$, that

$$\begin{aligned} |\mathcal{E}(\varphi+v) - \mathcal{E}(\varphi)| &= \left| \int_0^1 (1-t)\langle \mathcal{E}''(\varphi+tv)v, v\rangle \, dt \right| \\ &\le \int_0^1 (1-t)\|\mathcal{E}''(\varphi+tv)\| \cdot \|v\|^2 \, dt. \end{aligned}$$

It follows that

$$|\mathcal{E}(\varphi+v) - \mathcal{E}(\varphi)| \le (\gamma + \|L\|) \cdot \|v\|^2 \quad \forall v \in \mathcal{H}, \|v\| < \sigma. \tag{2.25d}$$

Combining (2.25c) with (2.25d), we find that

$$c \cdot |\mathcal{E}(\varphi+v) - \mathcal{E}(\varphi)|^{1/2} \le \|\mathcal{E}'(\varphi+v)\| \quad \forall v \in \mathcal{H}, \|v\| < \sigma,$$

where $c := \gamma/\sqrt{\gamma + \|L\|} < 1$. This completes the proof. □

In [**91**], J. W. Neuberger considers for a nonnegative C^1 functional $\mathcal{E} : U \to \mathbb{R}$ the following gradient inequality:

$$c \cdot \mathcal{E}(u)^{1/2} \le \|\mathcal{E}'(u)\|_{\mathcal{H}}, \quad u \in U,$$

where $c > 0$ is a constant. It follows that if $\varphi \in U$ is a critical point of $\mathcal{E}$, i.e., $\mathcal{E}'(\varphi) = 0$, then $\mathcal{E}(\varphi) = 0$. Hence, an easy argument yields that $\mathcal{E}'$ satisfies an optimal gradient inequality near every critical point of $\mathcal{E}$.

8. Remarks on gradient inequalities of type $\mathcal{G}_k$

Because of the importance of gradient inequalities of type $\mathcal{G}_k$ for the convergence results presented in the latter chapters, we present also some more useful properties of the function classes $\mathcal{G}_k$ below. Moreover, we will answer several questions that arise from practical uses of gradient inequalities.

(a) Let $\phi \in \mathcal{G}_1$. Then we have $0 \le \frac{1}{\phi} \in L^1_{loc}(\mathbb{R})$ and

$$\phi(x+y) \le \phi(x) + |y|^{1/2} \quad \forall x, y \in \mathbb{R}.$$

It follows that $\phi(x) \le \phi(x+y) + |-y|^{1/2}$ for all $x, y \in \mathbb{R}$ and thus

$$|\phi(x+y) - \phi(x)| \le |y|^{1/2} \quad \forall x, y \in \mathbb{R}.$$

In other words, each function of the class $\mathcal{G}_1$ is uniformly Hölder continuous with the exponent $\alpha = 1/2$. In notation, we have $\mathcal{G}_1 \subset C^{\frac{1}{2}}(\mathbb{R})$.

(b) Let $\phi \in \mathcal{G}$ and define a new function $\tilde{\phi}$ by

$$\tilde{\phi}(x) := \inf\{\phi(y) + |x-y|^{1/2} : y \in \mathbb{R}\} \quad (x \in \mathbb{R}). \tag{2.26a}$$

We have

$$0 \le \tilde{\phi} \le \phi. \tag{2.26b}$$

Moreover, for any $x, y \in \mathbb{R}$, we have

$$\begin{aligned} \tilde{\phi}(x+y) &= \inf\{\phi(z) + |x+y-z|^{1/2} : z \in \mathbb{R}\} \\ &\le \inf\{\phi(z) + |x-z|^{1/2} + |y|^{1/2} : z \in \mathbb{R}\} \\ &= \tilde{\phi}(x) + |y|^{1/2} \end{aligned}$$

and thus $\tilde{\phi} \in C^{\frac{1}{2}}(\mathbb{R})$. For the above inequality, we have used the subadditivity of the square root function: $|a+b|^{1/2} \le |a|^{1/2} + |b|^{1/2}$ for all $a, b \in \mathbb{R}$.

$\tilde{\phi}$ is the biggest function ψ in $C^{\frac{1}{2}}(\mathbb{R})$ such that $\psi \le \phi$. From this we claim that ϕ dominates a function of class $\mathcal{G}_1$ if and only if $\tilde{\phi} \in \mathcal{G}$.

In particular, we have the following assertions (i)–(ii):

(i) Each member in $\mathcal{G}_k$ dominates a member in $\mathcal{G}_1$.

(ii) If $\phi \in \mathcal{G}$ is even and quasi-monotone, i.e., $\phi(-x) = \phi(x)$ $(\forall x \in \mathbb{R})$ and if there exists a constant $c > 0$ such that

$$(*) \qquad \phi(y) \ge c \cdot \phi(x) \quad \forall 0 \le x \le y \le 2x,$$

then ϕ dominates a member in $\mathcal{G}_1$. Particularly, each even member in $\mathcal{G}$ whose restriction in the half-axis $\mathbb{R}_+$ is an increasing function dominates a member in $\mathcal{G}_1$.

To show (i), we assume that $\phi \in \mathcal{G}_k$ with some $k \ge 1$. Then we have

$$\phi(x+y) \le k\phi(x) + |y|^{1/2} \quad \forall x, y \in \mathbb{R}.$$

This implies that $k\tilde{\phi} \ge \phi$ and thus $\tilde{\phi} \in \mathcal{G}_1$ as desired.

To show (ii), we assume that $\phi \in \mathcal{G}$ is even and satisfies $(*)$. Fix $x \in \mathbb{R}$. For $y \in \mathbb{R}$, we consider two cases: either $|x-y| \ge |x|/3$ or $|x-y| \le |x|/3$.

For the first case, we have $\phi(|y|) + |x-y|^{1/2} \ge (|x|/3)^{1/2}$. For the second case, we have that $2|x|/3 \le |y| \le 4|x|/3$ and thus, by $(*)$,

$$\phi(y) + |x-y|^{1/2} = \phi(|y|) + |x-y|^{1/2} \ge \phi(|y|) \ge c \cdot \phi(2|x|/3).$$

Therefore, we have

$$\tilde{\phi}(x) \ge \min\{c \cdot \phi(2|x|/3), (|x|/3)^{1/2}\}$$

for all $x \in \mathbb{R}$. Clearly, this implies that $\tilde{\phi} \in \mathcal{G}$.

(c) The above result in **(b)** has the following consequence: If we have already had a gradient inequality with some function $\phi \in \mathcal{G}_k$ $(k \ge 1)$, then we can replace the function ϕ by a smaller function $\tilde{\phi} \in \mathcal{G}_1$ and thus we will have a gradient inequality of type $\mathcal{G}_1$. From this consequence one might consider the introduction of the general classes $\mathcal{G}_k$ to be superfluous. However, in the following chapters, we will continue using gradient inequalities of the general type $\mathcal{G}_k$, rather than using the special type $\mathcal{G}_1$. This is based on two reasons. First, in the practical manipulations, we usually have a gradient inequality of type $\mathcal{G}_k$ with $k > 1$ (e.g., a gradient inequality of LS-type that is of class $\mathcal{G}_2$). A more explicit form of the function involved in the gradient inequalities helps a lot in estimating the convergence rate. Second, the use of a gradient inequality of type $\mathcal{G}_1$ does not improve the convergence results in any sense, and moreover, the construction procedure for a dominated $\mathcal{G}_1$ function from a given $\mathcal{G}_k$ function is certainly complicated, as has been shown above.

(d) Does every member in $\mathcal{G}$ dominate a member in $\mathcal{G}_1$? The following construction gives a negative answer to this question.

EXAMPLE 8.1. Let $\{I_n := (a_n, b_n)\}_{n\geq 1}$ be a sequence of disjoint open subintervals of the unit interval $I := [0,1]$ such that the Lebesgue measure of the union $G := \bigcup_{n=1}^{\infty} I_n$ is 1. For the function

$$g(x) := x(x-1)(x^2+x-1) \quad (0 \leq x \leq 1), \tag{2.27a}$$

we have that $g'(x) = 1 - 4x + 4x^3$ and thus

$$g'(x) \leq 1 \quad \text{and} \quad g'(x) = 1 \quad \text{iff} \quad x \in \{0,1\}. \tag{2.27b}$$

We define a function $h : [0,1] \to \mathbb{R}$ by letting $h(x) = 0$ if $x \notin G$ and

$$h(x) := (b_n - a_n) \cdot g\left(\frac{x - a_n}{b_n - a_n}\right) \quad \forall x \in I_n = (a_n, b_n) \tag{2.28a}$$

for all $n \in \mathbb{N}$. Consider the function

$$f(x) := x - h(x) \quad (x \in I). \tag{2.28b}$$

It is easily seen that f is a C^1 function whose derivative $f'(x) = 1 - h'(x)$ is nonnegative and $f'(x) = 0$ iff $x \in \partial G$, by noting (2.27b). Since $h(0) = 0 = h(1)$ by noting that $g(0) = 0 = g(1)$, the range of f is the unit interval I. By Lemma 3.3 we claim that the function $\phi_f : I \to \mathbb{R}_+$ given by

$$\phi_f(f(x)) = f'(x) \quad \forall x \in [0,1] \tag{2.29}$$

is such that $0 \leq \frac{1}{\phi_f} \in L^1(I)$.

To find an estimate for the function ϕ_f, we consider a fixed subinterval I_n. Then $\phi_f(x - h(x)) = 1 - h'(x)$ $(\forall x \in I_n)$. Equivalently,

$$\phi_f(a_n + (b_n - a_n)(y - g(y))) = 1 - g'(y) \tag{2.30a}$$

for all $y := (x - a_n)/(b_n - a_n) \in I$. It follows from the equalities $y - g(y) = y^2(2 - y^2)$ and $1 - g'(y) = 4y(1-y)(1+y)$ that

$$4y(1-y) \leq \phi_f(a_n + (b_n - a_n)y^2(2-y^2)) \leq 8y(1-y) \tag{2.30b}$$

for all $y \in I$. Now an elementary manipulation yields that

$$\frac{c_1}{\sqrt{b_n - a_n}}\phi_n(x) \leq \phi_f(x) \leq \frac{c_2}{\sqrt{b_n - a_n}} \cdot \phi_n(x) \quad \forall x \in I_n \tag{2.31a}$$

where c_1, c_2 are two absolute constants and

$$\phi_n(x) := \min\{|x - a_n|^{1/2}, |b_n - x|^{1/2}\} \quad (x \in \mathbb{R}). \tag{2.31b}$$

We claim that the assertion that ϕ_f dominates a member in $\mathcal{G}_1$ is equivalent to the following convergence condition:

$$\sum_{n=1}^{\infty} \sqrt{b_n - a_n} < \infty. \tag{2.32}$$

To see this, we assume (2.32) and consider the function

$$\hat{\phi}_f(x) := \min\{\phi_n(x) : n \geq 1\} \quad (x \in \mathbb{R}). \tag{2.33}$$

We have that $c_1 \cdot \hat{\phi}_f(x) \leq \phi_f(x)$ for all $x \in I$ and thus, it suffices to show $\hat{\phi}_f \in \mathcal{G}_1$. Clearly, $\hat{\phi}_f$ satisfies the inequality $\hat{\phi}_f(x+y) \leq \hat{\phi}_f(x) + |y|^{1/2}$

$(\forall x, y \in \mathbb{R})$. Moreover, we have $\hat{\phi}_f(x) \geq \min\{|x|^{1/2}, |x-1|^{1/2}\}$ for all $x \notin I$ and for each $n \in \mathbb{N}$, $\hat{\phi}_f(x) = \phi_n(x) \quad (\forall x \in I_n)$. It follows that $\int_{I_n} \frac{dx}{\hat{\phi}_f(x)} = \sqrt{2(b_n - a_n)}$ for all $n \in \mathbb{N}$ and thus

$$\int_I \frac{dx}{\hat{\phi}_f(x)} = \sum_{n=1}^{\infty} \int_{I_n} \frac{dx}{\hat{\phi}_f(x)} = \sum_{n=1}^{\infty} \sqrt{2(b_n - a_n)},$$

which is finite by condition (2.32). Therefore, $\hat{\phi}_f \in \mathcal{G}_1$ as desired.

Conversely, we assume there exists some $\phi \in \mathcal{G}_1$ such that $\phi(x) \leq \phi_f(x)$ for all $x \in I$. Then, since $\phi_f(a_n) = \phi_f(b_n) = 0$ for all $n \in \mathbb{N}$, we have $\phi(a_n) = \phi(b_n) = 0$ for all $n \in \mathbb{N}$. Fix $n \in \mathbb{N}$ and consider $x \in I_n = (a_n, b_n)$. Then

$$0 \leq \phi(x) \leq \phi(a_n) + |x - a_n|^{1/2} = |x - a_n|^{1/2}$$

and $\phi(x) \leq |b_n - x|^{1/2}$. Therefore, $\phi(x) \leq \phi_n(x)$ for all $x \in I_n$ and thus

$$\sum_{n=1}^{\infty} \sqrt{2(b_n - a_n)} = \sum_{n=1}^{\infty} \int_{I_n} \frac{dx}{\phi_n(x)} \leq \int_I \frac{dx}{\phi(x)} < \infty.$$

This is merely condition (2.32).

The C^1 function f constructed in Example 8.1 has the property that the set $\mathrm{Crit}(f)$ of its critical values is simply the set ∂G, i.e.,

$$\mathrm{Crit}(f) = \partial G = I \setminus G.$$

How complicated could the critical values $I \setminus G$ of the C^1 function f be, if f' satisfies a gradient inequality of type $\mathcal{G}_1$? This is equivalent to asking how complicated could the set $I \setminus G$ be, if the sequence $\{(a_n, b_n)\}_{n \geq 1}$ satisfies condition (2.32)? The following set theoretic construction shows that there exists such an open subset G for which $I \setminus G$ is homeomorphic to the Cantor continuum.

EXAMPLE 8.2. Our construction method is a modification of the usual one for the Cantor continuum.

Let $\{\alpha_n\}_{n \geq 1}$ be a positive sequence such that

$$\sum_{n=1}^{\infty} 2^{n-1} \alpha_n = 1 \quad \text{and} \quad \sum_{n=1}^{\infty} 2^{n-1} \sqrt{\alpha_n} < \infty. \tag{2.34}$$

In the first step, we remove from the unit interval I an open interval whose length is α_1 and whose center is the middle point $1/2$. In the second step we remove from each of the remaining two closed intervals an open interval whose length is α_2 and whose center is the middle point of the corresponding interval. In general, in the nth step we remove from each of the remaining 2^{n-1} closed intervals an open interval whose length is α_n and whose center is the middle point of the corresponding interval. Finally, we let G be the union of the removed open subintervals. Then we have

$$\partial G = I \setminus G \cong C \quad \text{and} \quad \lambda_1(G) = \sum_{n=1}^{\infty} 2^{n-1}\alpha_n = 1.$$

Moreover, there exists an arrangement $\{I_n = (a_n, b_n)\}_{n\geq 1}$ of the subintervals of G such that

$$\sum_{n=1}^{\infty} \sqrt{b_n - a_n} = \sum_{n=1}^{\infty} 2^{n-1}\sqrt{\alpha_n} < \infty$$

by (2.34). By applying the result in Example 8.1 we obtain a C^1 function $f : I \to I$ such that $\phi(f(x)) \leq f'(x)$ ($x \in I$) with some $\phi \in \mathcal{G}_1$ and $\mathrm{Crit}(f) = \partial G \cong C$.

Concrete examples of sequences satisfying (2.34) can be found as follows. For $k \in \mathbb{Z}, k \geq 0$, we let

$$\rho_k(x) := \sum_{n=1}^{\infty} n^k x^n \quad (0 \leq x < 1).$$

These functions can be computed by the following recursion formula:

$$\rho_0(x) = x/(1-x), \quad \rho_{k+1}(x) = x\rho_k'(x) \quad (k \geq 1).$$

Fix $k \geq 2$. We let β_k, $0 < \beta_k < 1/2$, be the solution of the equation $\rho_k(x) = 2$ and set

(2.35) $$\alpha_n := n^k(\beta_k/2)^n \quad (n \in \mathbb{N}).$$

Then

$$\sum_{n=1}^{\infty} 2^{n-1}\alpha_n = 1, \quad \sum_{n=1}^{\infty} 2^{n-1}\sqrt{\alpha_n} = \frac{1}{2}\sum_{n=1}^{\infty} n^{k/2}(2\beta_k)^{n/2} < \infty,$$

since $0 < 2\beta_k < 1$. Therefore, the sequence given by (2.35) satisfies (2.34).

As seen in Example 8.1, the validity of a gradient inequality of type $\mathcal{G}_1$ hangs on the structure of the critical values set of the functional in question. Therefore, the worst-case scenario for the failure of a gradient inequality is the case where every value of the functional is critical. The following two-dimensional example shows that this can happen, if the function is only of C^1 class.

EXAMPLE 8.3. We choose the open set G in such a way that the boundary ∂G is just the Cantor continuum C. Using the method given in Example 8.1, we construct the C^1 function f that has the property that

$$\mathrm{Crit}(f) = C.$$

The C^1 function $V : I \times I \to [0, 2]$, given by

$$V(x, y) := f(x) + f(y) \quad (x, y \in I),$$

has the property that $\nabla V(x,y) = (f'(x), f'(y))$ and thus $\nabla V(x,y) = 0$ if and only if $(x,y) \in C \times C$. By the construction, we have $f(x) = x$ for all $x \in C$. Hence,

$$\mathrm{Crit}(V) = \{x + y : (x,y) \in C \times C\} = C + C.$$

Now a result of J. E. Neymann ("The sum of the Cantor set with itself", *Enseign. Math., II. Ser. 39*, No. 1-2, 177-178 (1993)) asserts that $C + C = [0,2]$. Thus, $\mathrm{Crit}(V) = [0,2]$. In other words, the C^1 function V does not have any regular value! This implies that the unique nonnegative function $\phi : [0,2] \to \mathbb{R}_+$ such that $\phi(V(x,y)) \le |V(x,y)|$ $(\forall (x,y) \in I \times I)$ is the trivial one $\phi \equiv 0$.

Recall that the famous Sard Lemma [**22**, Theorem 10.1, p. 55] asserts that the set of critical values of a C^r function defined on an open subset of $\mathbb{R}^n$ has Lebesgue measure zero if $r > n - 1$. Our two-dimensional example above shows that the C^2 smoothness requirement in Sard's Lemma for this case cannot be reduced into the C^1 smoothness.

(e) Should the structure of the set of critical points of a sufficiently smooth functional be "simple" in some sense if its gradient satisfies a gradient inequality of type $\mathcal{G}_1$? The answer to this question is "no". In fact, one can construct one-dimensional C^l ($l \in \mathbb{N}$ is arbitrary) functions that satisfy a gradient inequality of type $\mathcal{G}_1$, but the structure of its critical point set is allowed to be arbitrarily complicated. This reveals the advantage of using gradient inequalities of the general type $\mathcal{G}_k$ rather than using gradient inequalities of the special LS-type or weak LS-type, since the latter ones have forced the set of critical values to be locally a one-point set.

(f) We add some more comments on gradient inequalities of weak LS-type. For simplicity, we also consider the one-dimensional case. In this case, the derivative f' of a C^1 function $f : (-1,1) \to \mathbb{R}$ is said to satisfy a gradient inequality of weak LS-type near the critical point $x = 0$ if there holds

$$\phi(|f(x) - f(0)|) \le |f'(x)| \quad \forall |x| < \sigma$$

with a positive constant $\delta < 1$ and a function $\phi \in \mathcal{G}$ which is quasi-monotone in the sense that $\phi(y) \ge c\cdot\phi(x)$ for all $0 \le x \le y \le 2x$ with a constant $c > 0$. As is easily verified, the quasi-monotonicity of ϕ combined with the local integrability of $1/\phi$ implies that $\phi(x) > 0$ for all $x > 0$ and thus $f(0)$ is the unique critical value for the restriction of f in $I := (-\delta, \delta)$. This implies that the functions constructed in Example 8.3 do not satisfy a gradient inequality of weak LS-type near any critical point, since the set of critical values of such a function is a continuum.

The following assertion shows that the validity of gradient inequality of weak LS-type is more or less the reflection of the convexity of the function.

EXAMPLE 8.4. Let $f : [-1,1] \to \mathbb{R}$ be a strictly increasing C^1 function such that $f(0) = 0 = f'(0)$. Assume that f is convex, i.e., the function $f'(x)$

is nondecreasing. We define two functions $\phi_j : \mathbb{R}_+ \to \mathbb{R}$ $(j = 1, 2)$ by

$$\phi_1(f(x)) := f'(x) \quad x \in [0, 1],$$

$\phi_1(y) := \phi_1(f(1))$ for all $x > f(1)$, and

$$\phi_2(-f(x)) := -f'(x) \quad x \in [-1, 0],$$

$\phi_2(y) := \phi_1(-f(-1))$ for all $x > -f(-1)$. Both functions $\phi_{1,2}$ are nondecreasing by the convexity of f, and their reciprocals are locally integrable over $[0, \infty)$. Consider the function $\phi : \mathbb{R} \to \mathbb{R}_+$, given by

$$\phi(x) := \min\{\phi_1(|x|), \phi_2(|x|)\} \quad (x \in \mathbb{R}).$$

Clearly, $\phi \in \mathcal{G}$ and its restriction on $\mathbb{R}_+$ is a nondecreasing function. Moreover, there holds

$$\phi(|f(x)|) \leq |f'(x)| \quad \forall x \in [-1, 1].$$

Therefore, f' satisfies a gradient inequality of weak LS-type near the critical point $x = 0$.

In other words, Example 8.4 shows that the derivative of a convex function satisfies a gradient inequality of weak LS-type near any critical point of it. This assertion applies particularly to the following well known pathological function, given by

$$f(x) := \begin{cases} \exp(-1/x^2) & \text{if} \quad x \neq 0, \\ 0 & \text{if} \quad x = 0. \end{cases}$$

This function is of C^∞ class and all of its derivatives vanish at the point $x = 0$. Hence, its derivative f' does not satisfy a gradient inequality of LS-type near the critical point $x = 0$. However, a direct calculation shows that this function is convex on the interval $[-\sqrt{2/3}, \sqrt{2/3}]$ and thus near the critical point $x = 0$ its derivative satisfies a gradient inequality of weak LS-type.

(g) Last but not least, we would like to draw the reader to the following facts that will be seen in the latter chapters: A gradient inequality of type $\mathcal{G}_k$ is generally sufficient for the purpose of deriving convergence results from that gradient inequality. As discussed in the first chapter, if the gradient map $\mathcal{E}'$ has the property that it satisfies a gradient inequality near every critical point of it, then a compact solution to the autonomous gradient system $\dot{u} = -\mathcal{E}'(u)$ will converge to one single equilibrium. In many practical applications, it is hoped that this convergence property will be kept for perturbed systems $\dot{u} = -\mathcal{E}'(u) + g(t)$ with perturbations g that vanish as $t \to \infty$ in some manner, say $g(t) \sim t^{-\delta}$ as $t \to \infty$ with some $\delta > 0$. Now if we know that the type of gradient inequalities satisfied by the gradient map $\mathcal{E}'$ is of type $\mathcal{G}_k$, then we are able to claim that the above property of converging to one single equilibrium remains valid for compact solutions of the perturbed systems, once the decay exponent δ of the perturbations g is larger than $3/2$, i.e., $\delta > 3/2$. The difference between the validity of gradient inequalities of weak LS-type and type $\mathcal{G}_k$ is merely concerned with the decay rate of the perturbations. If the type of gradient inequalities is the weak LS-type, then the above convergence property remains valid if the decay exponent $\delta > 1$. The latter bound $\delta > 1$ is optimal, since $\delta > 1$ is also the necessary condition for the convergence of solutions to the one-dimensional system $\dot{u} = (1+t)^{-\delta}$. However, as shown before, if the set of critical values of the functional is not locally constant, then no gradient inequality of weak LS-type is available, although gradient inequalities of type $\mathcal{G}_k$ are valid.

CHAPTER 3

Abstract convergence results

1. Gradient-like systems

As in the previous chapter, the notation U is used below to denote an open subset of a real Banach space $(X, \|\cdot\|_X)$. Moreover, Y is a subspace of the dual space X', which becomes a Banach space under its own norm $\|\cdot\|_Y$. It follows that the embedding $(Y, \|\cdot\|_Y) \hookrightarrow (X', \|\cdot\|_{X'})$ is continuous and thus there exists a constant $\delta > 0$ such that

$$|\langle u, y\rangle_{X\times X'}| \leq \delta \cdot \|u\|_X \cdot \|y\|_Y \quad \forall (u, y) \in X \times Y. \tag{3.1a}$$

Here, $\langle \cdot, \cdot \rangle_{X\times X'}$ is the canonical bilinear form on $X \times X'$. Without loss of generality we may assume that $\delta = 1$ so that

$$\|y\|_{X'} \leq \|y\|_Y \quad \forall y \in Y. \tag{3.1b}$$

We use (3.1a) to identify X as a subset of the dual space of Y. It follows, by the assumption $\delta = 1$, that

$$\|u\|_{Y'} \leq \|u\|_X \quad \forall u \in X, \tag{3.1c}$$

where $\|\cdot\|_{Y'}$ is the usual norm on Y'. Note that we also have

$$\|u\|_{Y'} = \sup\{|\langle u, y\rangle_{X\times X'}| : y \in Y, \|y\|_Y \leq 1\} \quad \forall u \in X. \tag{3.1c$'$}$$

In general, we do not need to assume that $\|\cdot\|_{Y'}$ induces a norm on X.

We introduce a notion.

DEFINITION 1.1. By a *trajectory* in U, we mean a map $u : [0, T) \to U$ satisfying the following conditions (i)–(ii):

(i) (Strong continuity) The map $u : [0, T) \to U$ is continuous with respect to the norm of X.

(ii) (Weak-star differentiability) There exists a map $\dot{u} : [0, T) \to Y'$, called the *weak-star derivative* of u, such that for each fixed $y \in Y$ the function $[0, T) \ni t \mapsto \langle y, \dot{u}(t)\rangle_{Y\times Y'}$ is continuous and such that

$$\langle u(a) - u(b), y\rangle_{X\times X'} = \int_a^b \langle y, \dot{u}(t)\rangle_{Y\times Y'}\, dt \quad \forall a, b \in [0, T). \tag{3.1d}$$

In the sequel, we use

$$O(u) := \{u(t) : t \in [0, T)\}$$

to denote the *orbit* of the trajectory u.

The weak-star continuity of the derivative $\dot{u}$ implies for each compact $K \subset [0,T)$ that

$$\sup_{t\in K} |\langle y, \dot{u}(t)\rangle_{Y\times Y'}| < \infty \quad \forall y \in Y.$$

Thus, an application of the classical Resonance Theorem yields that the norms $\|\dot{u}(t)\|_{Y'}$ are uniformly bounded in K. Moreover, it follows that the norm function $t \mapsto \|\dot{u}(t)\|_{Y'}$ is lower semi-continuous.

PROPOSITION 1.2. *Let $\mathcal{M} = \mathcal{E}' : U \to Y$ be a gradient map associated with a C^1 functional $\mathcal{E} : U \to \mathbb{R}$. Assume $u : [0,T) \to U$ to be a trajectory in U. Then the function $t \mapsto \mathcal{E}(u(t))$ is continuously differentiable and*

$$\frac{d}{dt}\mathcal{E}(u(t)) = \langle \mathcal{M}(u(t)), \dot{u}(t)\rangle_{Y\times Y'} \quad \forall t \in [0,T).$$

PROOF. Fix $t_0 \in [0,T)$. Let $V \subset [0,T)$ be a neighborhood of t_0 such that all points $u(t)$ ($t \in V$) are contained in a convex subset of U; the existence of such V is guaranteed by the continuity of u, i.e., $\lim_{t\to t_0} \|u(t) - u(t_0)\|_X = 0$. Consider $t \in V$. Note that

$$\frac{d}{ds}\mathcal{E}(su(t_0) + (1-s)u(t)) = \mathcal{E}'(su(t_0) + (1-s)u(t))(u(t_0) - u(t))$$

for all $s \in [0,1]$. Using $\mathcal{E}' = \mathcal{M}$ and integrating the above representation, we find that

$$\begin{aligned}\mathcal{E}(u(t_0)) - \mathcal{E}(u(t)) &= \int_0^1 \langle u(t_0) - u(t), \mathcal{M}(su(t_0) + (1-s)u(t))\rangle_{X\times X'}\, ds \\ &= I_1 + I_2,\end{aligned}$$

where

$$I_1 := \langle u(t_0) - u(t), \mathcal{M}(u(t_0))\rangle_{X\times X'} \quad \text{and}$$

$$I_2 = \int_0^1 \langle u(t_0) - u(t), \mathcal{M}(su(t_0) + (1-s)u(t)) - \mathcal{M}(u(t_0))\rangle_{X\times X'}\, ds.$$

Applying (3.1d) to I_1, we find that $I_1 = \int_{t_0}^t \langle \mathcal{M}(u(t_0)), \dot{u}(\tau)\rangle_{Y\times Y'}\, d\tau$. It follows from the weak-star continuity of $\dot{u}$ that

$$\lim_{t\to t_0} I_1/(t - t_0) = \langle \mathcal{M}(u(t_0)), \dot{u}(t_0)\rangle_{Y\times Y'}. \tag{3.2}$$

On the other hand, we have

$$\begin{aligned}|I_2| &\le \int_0^1 \|u(t_0) - u(t)\|_{Y'} \cdot \|\mathcal{M}(su(t_0) + (1-s)u(t)) - \mathcal{M}(u(t_0))\|_Y\, ds \\ &\le M \cdot \|u(t_0) - u(t)\|_{Y'},\end{aligned}$$

where

$$M := \sup\{\|\mathcal{M}(su(t_0) + (1-s)u(t)) - \mathcal{M}(u(t_0))\|_Y : s \in [0,1]\}.$$

By the continuity of $\mathcal{M}$, we see that $M \to 0$ as $t \to t_0$. Moreover, using (3.1d) again, we find that

$$\|u(t_0) - u(t)\|_{Y'} \le \int_{\min\{t_0,t\}}^{\max\{t_0,t\}} \|\dot{u}(\tau)\|_{Y'} \, d\tau.$$

As seen before, the integrand $\|\dot{u}(\tau)\|_{Y'}$ is locally uniformly bounded. Taking these into account, we conclude that $I_2/(t-t_0) \to 0$ as $t \to t_0$. Therefore, $\mathcal{E}(u(t))$ is differentiable at the point t_0 and (see (3.2))

$$\begin{aligned} \lim_{t\to t_0} (t-t_0)^{-1}[\mathcal{E}(u(t)) - \mathcal{E}(u(t_0))] &= \lim_{t\to t_0} (t-t_0)^{-1} I_1 \\ &= \langle \mathcal{M}(u(t_0)), \dot{u}(t_0)\rangle_{Y\times Y'}. \end{aligned}$$

This completes the proof. □

Now let $\mathcal{M} = \mathcal{E}' : U \subset X \to Y$ be a gradient map. It follows from Definition 1.1 of Chapter 2 that $\mathcal{M}$ is a continuous map and the potential $\mathcal{E} : X \to \mathbb{R}$ is of class C^1. Consider a continuous map $J : Y \to Y'$. Then the product $J\mathcal{M}$ is a continuous map from U into Y'. In many applications, one needs to solve the following differential equation:

$$\begin{cases} \dot{u}(t) = J\mathcal{M}(u(t)), & t \ge 0, \\ u(0) = u_0 \in U. \end{cases} \tag{3.3a}$$

One usual way of solving (3.3a) is to solve its integral form

$$u(t) = u_0 + \int_0^t J\mathcal{M}(u(s))\, ds, \quad t \in [0,T). \tag{3.3b}$$

There is a weaker variant of (3.3b), namely

$$\langle u(t) - u_0, y\rangle_{X\times X'} = \int_0^t \langle y, J\mathcal{M}(u(s))\rangle_{Y\times Y'}\, ds \quad \forall y \in Y; t \in [0,T). \tag{3.3c}$$

DEFINITION 1.3. (i) A continuous map $u : [0,T) \to U$ is called a *mild solution* of (3.3a) if u is a solution of the integral equation (3.3b).

(ii) A map $u : [0,T) \to U$ is called a *weak solution* of (3.3a) if u is a trajectory in U (cf. Def. 1.1) and satisfies (3.4c), i.e.,

$$\langle y, \dot{u}(t) - J\mathcal{M}(u(t))\rangle_{Y\times Y'} = 0 \quad \forall y \in Y;\ t \in [0,T). \tag{3.4}$$

Clearly, a mild solution is also a weak solution.

Consider a weak solution $u : [0,T) \to U$ of (3.3a). It follows from Proposition 1.2 that the function $t \mapsto \mathcal{E}(u(t))$ is differentiable and, by (3.4), there holds

$$\frac{d}{dt}\mathcal{E}(u(t)) = \langle \mathcal{M}(u(t)), \dot{u}(t)\rangle_{Y\times Y'} = \langle \mathcal{M}(u(t)), J\mathcal{M}(u(t))\rangle_{Y\times Y'} \tag{3.5}$$

for all $t \in [0,T]$. The function $\mathcal{E}(u(t))$ is called the **energy** of the solution u. We see from (3.5) that the growth of the energy is determined completely by the operator J. Hence, we say that (3.3a) is a **dissipative system** if the operator J is dissipative in the sense that

$$\langle y, Jy\rangle_{Y\times Y'} \leq 0 \quad \forall y \in Y.$$

This implies that the energy of a solution of (3.3a) is a nonincreasing function. In many applications the map J is strictly dissipative in the sense that $\langle y, Jy\rangle_{Y\times Y'} < 0$ for all $0 \neq y \in Y$. One especially interesting case is when J is a strongly dissipative map such that

$$\|Jy\|_{Y'} \leq (1/c)\cdot\|y\|_Y \quad \text{and} \quad \langle y, Jy\rangle_{Y\times Y'} \leq -c\cdot\|y\|_Y^2 \quad (\forall y \in Y) \tag{3.6a}$$

for some positive constant $c > 0$. Typical examples of such maps are taken from the case where the dual Y' of the Banach space Y is uniformly strictly convex and $-J$ is the duality map of Y [**32**, Prop. 12.3, p. 115].

Under the above condition (3.6a) on the map J, we consider a solution u of (3.3a). Then we have

$$\begin{cases} \langle -\mathcal{M}(u(t)), \dot{u}(t)\rangle_{Y\times Y'} \geq c\cdot\|\mathcal{M}(u(t))\|_Y^2, \\ (1/c)\cdot\|\mathcal{M}(u(t))\|_Y \geq \|\dot{u}(t)\|_{Y'} \geq c\cdot\|\mathcal{M}(u(t))\|_Y. \end{cases} \tag{3.6b}$$

To compare solutions satisfying (3.6b) with the notion of *pseudo-gradient trajectory* defined by Palais in 1966 and widely used to construct solutions of variational problems (cf., e.g., Struwe [**112**] and Willem [**122**]), we consider $\mathcal{M}$ as the *gradient* of the C^1 functional $\mathcal{E}$. Then, a *pseudo-gradient vector field* for $\mathcal{M}$ is a locally continuous map $\mathcal{N} : U \to X$ such that, for every $u \in U$,

$$\begin{cases} \langle \mathcal{M}(u), \mathcal{N}(u)\rangle_{X'\times X} \geq \|\mathcal{M}(u)\|_{X'}^2, \\ 2\|\mathcal{M}(u)\|_{X'} \geq \|\mathcal{N}(u)\|_X \geq \|\mathcal{M}(u)\|_{X'}. \end{cases} \tag{3.7a}$$

A pseudo-gradient vector field is called to be a *gradient vector field* if the second inequality in (3.7a) is replaced by the equality $\|\mathcal{N}(u)\|_X = \|\mathcal{M}(u)\|_{X'}$ ($\forall u \in U$.)

Let $\mathcal{N}$ be a pseudo-gradient vector field for $\mathcal{M}$. Then solutions of the following equation

$$\begin{cases} \dot{v}(t) = -\mathcal{N}(v(t)), \quad t \geq 0, \\ v(0) = v_0 \in U \end{cases} \tag{3.7b}$$

are called pseudo-gradient trajectories in U. Thus, if $v : [0, T) \to U$ is a pseudo-gradient trajectory on U, then

$$\begin{cases} \langle -\mathcal{M}(v(t)), \dot{v}(t)\rangle_{Y\times Y'} \geq \|\mathcal{M}(v(t))\|_{X'}^2, \\ 2\|\mathcal{M}(v(t))\|_{X'} \geq \|\dot{v}(t)\|_X \geq \|\mathcal{M}(v(t))\|_{X'}. \end{cases} \tag{3.7c}$$

The similarity of (3.7c) and (3.6b) is easily seen. Therefore, we will call solutions of a strongly dissipative system *gradient-like trajectories*, and in this sense the corresponding system is called a *gradient-like system*.

The counterpart to strong dissipation is dispersion. We say that (3.3a) is a **dispersive system** if J is a dispersive map in the sense that $\langle y, Jy\rangle_{Y\times Y'} = 0$ ($\forall y \in Y$). This implies that (3.3a) is an energy conservation system, i.e., $\mathcal{E}(u(t))$ keeps constant for every solution of (3.3a). In the classical literature, a dispersive system is also called a **Hamilton system**.

Literature on the existence and regularity theory of nonlinear evolution equations is extensive, see, e.g., [**14, 123, 77, 63, 94, 24, 51, 32, 35**]. One especially useful reference is the recent monograph of R. E. Showalter [**105**], which covers almost all well known results for monotone and accretive operators.

2. Three technical lemmas

Before discussing convergence results in gradient-like systems, we present three elementary lemmas in this section, which will be used frequently in the sequel.

LEMMA 2.1. *Let $H : [t_0, t_1] \to \infty$ be a monotone and absolutely continuous function (hence $H' \in L^1([t_0, t_1])$). Let $\Phi : [H(t_1), H(t_0)] \to \mathbb{R}$ be an absolutely continuous function. Then the composition $G := \Phi \circ H$ defined by $G(t) := \Phi(H(t))$ ($t \in [t_0, t_1]$) is also absolutely continuous and there holds*

$$G'(t) = \Phi'(H(t))H'(t) \quad a.e. \quad t \in [t_0, t_1].$$

PROOF. Let $\varepsilon > 0$. On the one hand, we use the absolute continuity of Φ to find some positive constant δ such that

$$\sum_{j=1}^{n} |\Phi(d_j) - \Phi(c_j)| < \varepsilon$$

for any finitely many points $c_j, d_j \in [H(t_0), H(t_1)]$ with

$$(*) \qquad c_1 \le d_1 \le c_2 \le d_2 \le \cdots \le c_n \le d_n \quad \text{and} \quad \sum_{j=1}^{n} |d_j - c_j| < \delta.$$

On the other hand, since H is absolutely continuous, there exists another positive constant δ' such that if a_j, b_j ($j = 1, 2, ..., n$) are finitely many points in $[t_0, t_1]$ with

$$(**) \qquad a_1 \le b_1 \le a_2 \le b_2 \le \cdots \le a_n \le b_n \quad \text{and} \quad \sum_{j=1}^{n} |b_j - a_j| < \delta',$$

then there holds that

$$\sum_{j=1}^{n} |H(b_j) - H(a_j)| < \delta.$$

To show the absolute continuity of $\Phi \circ H$, we consider $2n$ points a_j, b_j with the afore mentioned property in $(**)$. Then we have either that

$$H(b_n) \le H(a_n) \le \cdots \le H(b_2) \le H(a_2) \le H(b_1) \le H(a_1)$$

if H is nonincreasing, or

$$H(b_n) \geq H(a_n) \geq \cdots \geq H(b_2) \geq H(a_2) \geq H(b_1) \geq H(a_1)$$

if H is nondecreasing. Hence the $2n$ points $H(b_{n-j}), H(a_{n-j})$ $(0 \leq j \leq n-1)$ satisfy the property in $(*)$. It follows that

$$\sum_{j=1}^{n} |\Phi(H(b_j)) - \Phi(H(a_j))| < \varepsilon.$$

This proves the absolute continuity of the composition $G = \Phi \circ H$. The equality $G' = (\Phi' \circ H)H'$ follows from the absolute continuity of G. $\square$

We make a remark concerning the result above. Recall the following equivalent description of absolute continuity: A function f is absolutely continuous if and only if it is continuous, of bounded variation, and has the property that $\lambda_1(f(E)) = 0$ whenever $\lambda_1(E) = 0$. Here λ_1 is the Lebesgue measure on the real line. In general, if both f and g are absolutely continuous, then their composition $f \circ g$ need not be absolutely continuous. Lemma 2.1 says that the composition $f \circ g$ is again absolutely continuous whenever g is monotone. The fact that prevents $f \circ g$ from being absolutely continuous is that $f \circ g$ need not be of bounded variation. A classical result of de la Vallée Poussin says that $f \circ g$ is absolutely continuous if and only if $f' \circ g \cdot g'$ is integrable.

LEMMA 2.2. *Let $h : \mathbb{R}_+ \to \mathbb{R}$ be a bounded, monotone and absolutely continuous function. Assume that there holds the inequality*

$$\phi(h(t)) \leq \sqrt{|h'(t)|} + g(t) \quad (\forall t \in \Lambda), \tag{3.8}$$

where Λ is a measurable subset of $\mathbb{R}_+$, $0 \leq g \in L^1(\mathbb{R}_+)$ and $\phi \in \mathcal{G}$ (i.e., $0 \leq \frac{1}{\phi} \in L^1_{loc}(\mathbb{R})$). Then $\sqrt{|h'|} \in L^1(\Lambda)$ and

$$\int_\Lambda \sqrt{|h'(t)|}\, dt \leq \|g\|_{L^1} + \left| \int_{h(0)}^{h(\infty)} \frac{ds}{\phi(s)} \right|.$$

Here $h(\infty) := \lim_{t\to\infty} h(t)$.

As a consequence, if $0 \leq f \in L^2(\mathbb{R}_+)$ and $0 \leq g \in L^1(\mathbb{R}_+)$ are such that

$$\phi\left(\int_t^\infty f(s)^2\, ds \right) \leq f(t) + g(t) \quad (\forall t \in \Lambda), \tag{3.8$'$}$$

where Λ is a measurable subset of $\mathbb{R}_+$ and $\phi \in \mathcal{G}$, then $f \in L^1(\Lambda)$ and

$$\int_\Lambda f(t)\, dt \leq \|g\|_{L^1} + \int_0^{\|f\|^2_{L^2}} \frac{ds}{\phi(s)}.$$

PROOF. We consider the case where h is nondecreasing, i.e., $h' \geq 0$. The proof for the other case, $h' \leq 0$, is completely similar and thus will be omitted.

Since $1/\phi \in L^1_{loc}(\mathbb{R})$, the function

$$\Phi(x) := \int_0^x \frac{ds}{\phi(s)} \quad (x \in \mathbb{R})$$

is absolutely continuous and $\Phi'(x) = 1/\phi(x)$ for a.e. $x \in \mathbb{R}$. It follows from Lemma 2.1 that the composition $\Phi \circ h$ is also absolutely continuous and there holds

$$\frac{d}{dt}\Phi(h(t)) = h'(t)/\phi(h(t)) \geq 0 \quad \text{a.e.} \quad t \in [0, \infty).$$

Using condition (3.8), we have for a.e. $t \in \Lambda$ that

$$h'(t)/\phi(h(t)) \geq h'(t)/(\sqrt{|h'(t)|} + g(t)) \geq \sqrt{|h'(t)|} - g(t)$$

and thus

$$\sqrt{|h'(t)|} \leq g(t) + \frac{d}{dt}\Phi(h(t)) \quad \text{a.e.} \quad t \in \Lambda.$$

Taking integration and noting that $g \geq 0$, we obtain

$$\int_\Lambda \sqrt{|h'(t)|}\, dt \leq \|g\|_{L^1} + \lim_{t\to\infty} \Phi(h(t)) - \Phi(h(0)).$$

This is the desired result, since $\Phi(h(t)) \to \Phi(h(\infty))$ as $t \to \infty$.

The remaining part of Lemma 2.2 corresponds to the special choice $h(t) := \int_t^\infty f(s)^2\, ds \quad (t \geq 0)$. □

We remark that a related result to Lemma 2.2, for the special case when $\phi(s) = |s|^\theta$ ($\theta \in (0,1)$) and $g \equiv 0$, has been proved in [**40**, Lemma 7.1]; see also [**70**, Lemma 4.1].

The following lemma will be used frequently in our subsequent proofs for convergence results. Recall that $\varphi \in X$ is said to be a *cluster point* of the orbit $O(u) = \{u(t) : t \geq 0\}$ of a trajectory $u : \mathbb{R}_+ \to X$ if there exists an unbounded sequence $\{t_n\}_{n\geq 1}$ such that

$$\|u(t_n) - \varphi\|_X \to 0 \quad \text{as} \quad n \to \infty.$$

LEMMA 2.3. *Let $u : \mathbb{R}_+ \to X$ be a trajectory (in the sense of Definition 1.1) whose orbit $O(u) = \{u(t) : t \geq 0\}$ is precompact in X. Moreover, assume that there exist a cluster point $\varphi \in X$ of $O(u)$, a constant $\sigma > 0$, and a continuous and injective map $\gamma : Y' \to Y'$ such that*

(3.9a) $$\int_W \|\gamma\dot{u}(t)\|_{Y'}\, dt < \infty,$$

where

(3.9b) $$W := \{t \geq 0 : \|u(t) - \varphi\|_X < \sigma\}.$$

Then the trajectory u converges as $t \to \infty$ to φ in the sense that

(3.9c) $$\lim_{t\to\infty} \|u(t) - \varphi\|_X = 0 \quad \textit{and} \quad \int_0^\infty \|\gamma\dot{u}(t)\|_{Y'}\, dt < \infty.$$

PROOF. By the continuity of the trajectory $u : \mathbb{R}_+ \to X$, we see that the set W given in (3.9b) is open. It is also non-empty, since φ is a cluster point of $O(u)$. We choose an unbounded sequence $(t_n) \subset W$ such that

$$(*) \qquad \|u(t_n) - \varphi\|_X \to 0 \quad \text{as} \quad n \to \infty.$$

For each t_n we let

$$\tilde{t}_n := \sup\{t \geq t_n : [t_n, t] \subset W\}.$$

We claim that there exists some $N \in \mathbb{N}$ such that $\tilde{t}_N = \infty$.

We give an argument by contradiction. Suppose for each $n \in \mathbb{N}$ we have $\tilde{t}_n < \infty$. Then we have that $[t_n, \tilde{t}_n) \subset W$, but the point $\tilde{t}_n$ is located in the boundary of W, i.e.,

$$(**) \qquad \|u(\tilde{t}_n) - \varphi\|_X = \sigma \quad \forall n \in \mathbb{N}.$$

Using (3.1d), we have that

$$\|\gamma u(\tilde{t}_n) - \gamma u(t_n)\|_{Y'} \leq \int_{t_n}^{\tilde{t}_n} \|\gamma \dot{u}(t)\|_{Y'}\, dt$$

and thus

$$\|\gamma u(\tilde{t}_n) - \gamma\varphi\|_{Y'} \leq \|\gamma u(t_n) - \gamma\varphi\|_{Y'} + \int_{t_n}^{\tilde{t}_n} \|\gamma \dot{u}(t)\|_{Y'}\, dt =: M_n$$

for all $n \in \mathbb{N}$. By (3.9a) and $(*)$, combined with the continuity of the map $\gamma : Y' \to Y'$, we have the convergence

$$M_n \to 0 \quad \text{as} \quad n \to \infty.$$

Consequently,

$$\lim_{n\to\infty} \|\gamma u(\tilde{t}_n) - \gamma\varphi\|_{Y'} = 0.$$

This implies that if $\tilde{\varphi} \in X$ is any cluster point of the precompact sequence $\{u(\tilde{t}_n) : n \in \mathbb{N}\} \subset O(u)$, then we must have $\|\gamma\tilde{\varphi} - \gamma\varphi\|_{Y'} = 0$ and thus $\tilde{\varphi} = \varphi$ by the injectivity of the map γ. Hence, φ is the unique cluster point of the precompact sequence $\{u(\tilde{t}_n) : n \in \mathbb{N}\}$ in X and thus $\|u(\tilde{t}_n) - \varphi\|_X \to 0$ as $n \to \infty$, contradicting $(**)$.

Therefore, we have proved that there exists some $N \in \mathbb{N}$ such that $\tilde{t}_N = \infty$, i.e., $[t_N, \infty) \subset W$. This implies by (3.9a) the integrability assertion in (3.9c). As a consequence of it, we have the convergence $\|\gamma\dot{u}(t) - \gamma\varphi\|_{Y'} \to 0$ as $t \to \infty$. A similar argument as before (by using the injectivity of the continuous map γ) yields that φ is the unique cluster point of the precompact orbit $O(u)$ in X and thus the convergence $u(t) \to \varphi$ as $t \to \infty$ in X follows. □

3. Convergence in gradient-like systems

Our setting in this section is the same as it was at the beginning of this chapter, i.e., U is an open subset of a Banach space X, and $\mathcal{E} : U \to \mathbb{R}$ is a C^1 functional for which there exists a Banach space $Y \subset X'$ such that $\mathcal{E}'(U) \subset Y$ and the gradient map $\mathcal{E}' : U \to Y$ is continuous.

Our task below is to study the convergence of **gradient-like trajectories** $u : [0, \infty) \to U$, with respect to the gradient map $\mathcal{E}' : U \to Y$, which locate themselves in the neighborhood of the set of stationary points

$$\mathcal{S} = \{\varphi \in U : \mathcal{E}'(\varphi) = 0\}.$$

Moreover, these trajectories satisfy a growth condition of the form:

$$\text{(3.10a)} \qquad \begin{cases} \langle -\mathcal{E}'(u(t)), \dot{u}(t)\rangle_{Y\times Y'} \geq \|\mathcal{E}'(u(t))\|_Y \cdot \|\gamma\dot{u}(t)\|_{Y'} + F'(t), \\ \|\mathcal{E}'(u(t))\|_Y \geq \|\gamma\dot{u}(t)\|_{Y'} + G'(t) \end{cases}$$

for all $t \geq 0$, where $\gamma : Y' \to Y'$ is an *injective* and continuous map (which could be *nonlinear*), and both $F, G : \mathbb{R}_+ \to \mathbb{R}_+$ are nonincreasing absolutely continuous functions satisfying

$$\text{(3.10b)} \qquad \lim_{t\to+\infty} (F(t) + G(t)) = 0 \quad \text{and} \quad \int_0^\infty \sqrt{F(t)}\, dt < \infty.$$

A simplified version of the growth condition (3.10a,b) is obtained by switching F, G into zero and weakening the inequality in the second line:

$$\text{(3.10}'\text{)} \qquad \begin{cases} \langle -\mathcal{E}'(u(t)), \dot{u}(t)\rangle_{Y\times Y'} \geq \|\mathcal{E}'(u(t))\|_Y \cdot \|\gamma\dot{u}(t)\|_{Y'}, \\ \mathcal{E}'(u(t)) = 0 \implies \dot{u}(t) = 0 \end{cases}$$

for all $t \geq 0$, where $\gamma : Y' \to Y'$ is again an *injective* and continuous map. By identifying X as a subspace of Y' (we have done so from the very beginning), we also identify γ as a continuous map from X to Y'. This identification will be used later without further explanation.

Note that the map γ and the functions F, G involved in (3.10a,b) and (3.10′) might be dependent on the given trajectory u.

Recall from the definition of trajectories that we have the low semicontinuity of the norm function $t \mapsto \|\gamma\dot{u}(t)\|_{Y'}$. Therefore, the Lebesgue integral

$$\text{(3.11a)} \qquad I_\gamma(u) := \int_0^\infty \|\gamma\dot{u}(t)\|_{Y'}\, dt$$

is well-defined. As in the finite-dimensional case, $I_\gamma(u)$ should be understood as the *length* of the curve $\gamma u : [0, \infty) \to U$ evaluated in the norm $\|\cdot\|_{Y'}$. To see the convergence implication of the finiteness of the integral $I_\gamma(u)$, we use (3.1d) to find that

$$\text{(3.11b)} \qquad \|\gamma u(a) - \gamma u(b)\|_{Y'} \leq \int_a^b \|\gamma\dot{u}(t)\|_{Y'}\, dt \quad (0 \leq a \leq b).$$

Therefore, the finiteness of the integral $I_\gamma(u)$ implies that the curve γu in Y' converges with respect to the norm $\|\cdot\|_{Y'}$. Our first result connects the finiteness of the integral $I_\gamma(u)$ with a certain gradient inequality.

PROPOSITION 3.1. *Let $u : [0,\infty) \to U$ be a trajectory and assume that $\gamma : Y' \to Y'$ is a bounded linear operator. Then the finiteness $I_\gamma(u) < \infty$ implies that the gradient map $\mathcal{E}' : U \to Y$ satisfies a gradient inequality in the orbit $O(u)$. More precisely, if*

$$I_\gamma(u) = \int_0^\infty \|\gamma\dot{u}(t)\|_{Y'}\, dt < \infty, \tag{3.12}$$

then there exists some ϕ with $0 \le \frac{1}{\phi} \in L^1(\mathbb{R})$, such that

$$\phi(\mathcal{E}(u(t))) \le \|\mathcal{E}'(u(t))\|_\gamma \quad \forall t \ge 0, \tag{3.13a}$$

where $\|\cdot\|_\gamma$ is defined as

$$\|w\|_\gamma := \sup\{|\langle w, v\rangle_{Y\times Y'}| : v \in Y', \|\gamma v\|_{Y'} \le 1\} \quad (w \in Y). \tag{3.13b}$$

PROOF. Consider the function

$$f(t) := \mathcal{E}(u(t)) \quad (t \ge 0).$$

It follows from Proposition 1.2 that f is a C^1 function and

$$\frac{d}{dt}\mathcal{E}(u(t)) = \langle \mathcal{E}'(u(t)), \dot{u}(t)\rangle_{Y\times Y'}, \quad t \ge 0.$$

Now we use the convergence $I_\gamma(u) < \infty$. Note that the lower semicontinuous function $t \mapsto \|\gamma\dot{u}(t)\|_{Y'}$ is locally bounded. Therefore, we can construct a strictly decreasing and bounded C^1 function g on $[0,\infty)$ such that $g(t) \to 0$ as $t \to \infty$ and $-g'(t) \ge \|\gamma\dot{u}(t)\|_{Y'}$ for all $t \ge 0$. The unique inverse $h := g^{-1} : (0, g(0)] \to [0,\infty)$ of g is also a C^1 function.

To find a function ϕ satisfying the desired gradient inequality, we consider the function $\Gamma : (0, g(0)] \to \mathbb{R}$ given by

$$\Gamma(x) := f(h(x)), \quad x \in (0, g(0)].$$

Γ is a C^1 function on the bounded interval $V := (0, g(0)]$. Therefore, an application of Theorem 3.2 of Chapter 2 to the pair (V, Γ) yields a function $\phi \in \mathcal{G}$ such that

$$\phi(\Gamma(x)) \le |\Gamma'(x)| \quad \forall x \in (0, g(0)].$$

We estimate the derivative of Γ as follows:

$$\begin{aligned} |\Gamma'(x)| &= |f'(h(x))h'(x)| = |\langle \mathcal{E}'(u(h(x))), \dot{u}(h(x))\rangle_{Y\times Y'} \cdot h'(x)| \\ &\le \|\mathcal{E}'(u(h(x)))\|_\gamma \cdot \left(\|\gamma\dot{u}(h(x))\|_{Y'} \cdot |h'(x)|\right). \end{aligned}$$

Here for the inequality we have used the definition of $\|\cdot\|_\gamma$ given in (3.13b) which is equivalent to the property that $|\langle w, v\rangle_{Y\times Y'}| \le \|w\|_\gamma \cdot \|\gamma v\|_{Y'}$ for all $(w, v) \in Y \times Y'$. Note that $\|\gamma\dot{u}(h(x))\|_{Y'} \le -g'(h(x))$. Therefore,

$$\begin{aligned} &\|\gamma\dot{u}(h(x))\|_{Y'} \cdot |h'(x)| \le |g'(h(x)) \cdot h'(x)| \\ &= |\frac{d}{dx}g(h(x))| = |\frac{d}{dx}x| = 1, \end{aligned}$$

since h is the inverse of g. Consequently, $|\Gamma'(x)| \le \|\mathcal{E}'(u(h(x)))\|_\gamma$. For $t \ge 0$ we take $x := g(t) \in (0, g(0)]$. Then we have that $t = h(x)$ and thus

$$\begin{aligned}\phi(\mathcal{E}(u(t))) &= \phi(f(t)) = \phi(f(h(x))) = \phi(\Gamma(x)) \\ &\le |\Gamma'(x)| \le \|\mathcal{E}'(u(h(x)))\|_\gamma = \|\mathcal{E}'(u(t))\|_\gamma.\end{aligned}$$

This is the desired inequality in (3.13a). □

It is worth noting that in Proposition 3.1, no relation between the gradient map $\mathcal{E}' : U \to Y$ and the given trajectory u is needed. This, combined with the fact that the class $\mathcal{G}$ is closed under finite minimization, implies the following consequence for any gradient map $\mathcal{E}' : U \to Y$: If $V = \cup_{j=1}^n O(u_j)$ is the union of finitely many trajectories $u_j : [0, \infty) \to U$ whose total length $\sum_{j=1}^n \int_0^\infty \|\dot{u}(t)\|_{Y'}\, dt$ evaluated in the Y' norm is finite, then for any gradient map $\mathcal{E}' : U \to Y$ there exists a function $\phi \in \mathcal{G}$ such that

$$\phi(\mathcal{E}(v)) \le \|\mathcal{E}'(v)\|_Y \quad \forall v \in V.$$

In other words, any gradient map $\mathcal{E}' : U \to Y$ satisfies a gradient inequality on every such subset V. Note that a subset V of the form above does not need to be precompact in the X norm topology.

Now we continue our study. We note that one counterpart to Proposition 3.1 is the following inverse problem: Does the validity of a gradient inequality imply the convergence of the integral $I_\gamma(u)$? First we consider trajectories satisfying the simplified growth condition (3.10′). For this case, we have an affirmative answer.

PROPOSITION 3.2. **(Weak convergence)** *Let $u : [0, \infty) \to U$ be a trajectory satisfying* (3.10′) *such that*

$$\sup\{|\mathcal{E}(u(t))| : t \ge 0\} < \infty. \tag{3.14}$$

If the gradient map $\mathcal{E}' : U \to Y$ satisfies a gradient inequality in the orbit $O(u)$, i.e., if there exists some ϕ with $0 \le \frac{1}{\phi} \in L^1_{loc}(\mathbb{R})$ such that

$$\phi(\mathcal{E}(u(t))) \le \|\mathcal{E}'(u(t))\|_Y \quad \forall t \ge 0, \tag{3.15a}$$

then

$$I_\gamma(u) = \int_0^\infty \|\gamma \dot{u}(t)\|_{Y'}\, dt \le \int_a^{\mathcal{E}(u(0))} \frac{ds}{\phi(s)} < \infty, \tag{3.15b}$$

where $a := \lim_{t\to\infty} \mathcal{E}(u(t))$.

PROOF. We consider the function $H(t) := \mathcal{E}(u(t))$ $(t \ge 0)$, which is a C^1 function with derivatives

$$-\dot{H}(t) = \langle -\mathcal{E}'(u(t)), \dot{u}(t)\rangle_{Y\times Y'}.$$

It follows from the growth condition (3.10′) that

$$-\dot{H}(t) \ge \|\mathcal{E}'(u(t))\|_Y \cdot \|\gamma \dot{u}(t)\|_{Y'}. \tag{3.16a}$$

Thus the C^1 function H is nonincreasing and uniformly bounded by (3.14).

Let Φ be the absolutely continuous function given by

$$\Phi(x) := \int_a^x \frac{1}{\phi(s)}\,ds \quad (x \in \mathbb{R}).$$

Here $a = \lim_{t\to\infty} H(t) > -\infty$ as in (3.15b). Φ is a.e. differentiable and $\Phi' = 1/\phi$. Moreover, by Lemma 2.1, we see that the composition $\Phi \circ H$ is absolutely continuous and there holds

$$\frac{d}{dt}\Phi(H(t)) = H'(t)/\phi(H(t)) \quad \forall t \in \Lambda, \tag{3.16b}$$

where $\Lambda \subset [0,\infty)$ is such that the complement $[0,\infty)\setminus\Lambda$ has Lebesgue measure $\lambda_1([0,\infty)\setminus\Lambda) = 0$.

Now for any $t \in \Lambda$ we have two possibilities: Either (i) $\|\gamma\dot{u}(t)\|_{Y'} = 0$ or (ii) $\|\gamma\dot{u}(t)\|_{Y'} > 0$ and thus $\|\mathcal{E}'(u(t))\|_Y > 0$ by the second line in condition (3.10′) combined with the fact that γ is injective. For case (ii) we use the gradient inequality (3.15a) to find that

$$\phi(H(t)) = \phi(\mathcal{E}(u(t))) \le \|\mathcal{E}'(u(t))\|_Y. \tag{3.16c}$$

This implies by (3.16a,b) that

$$-\frac{d}{dt}\Phi(H(t)) \ge \frac{\|\mathcal{E}'(u(t))\|_Y \cdot \|\gamma\dot{u}(t)\|_{Y'}}{\|\mathcal{E}'(u(t))\|_Y} = \|\gamma\dot{u}(t)\|_{Y'}. \tag{3.16d}$$

Therefore, by the positivity of the function $-\frac{d}{dt}\Phi(H(t))$, combined with the fact that $\lambda_1([0,\infty)\setminus\Lambda) = 0$, we finally obtain the following estimate:

$$-\frac{d}{dt}\Phi(H(t)) \ge \|\gamma\dot{u}(t)\|_{Y'} \quad \text{a.e.} \quad t \in [0,\infty)$$

for both cases (i) and (ii), which yields by integration that

$$\int_0^\infty \|\gamma\dot{u}(t)\|_{Y'}\,dt \le \Phi(H(0)).$$

This is (3.15b). □

Our first convergence theorem reads as follows.

THEOREM 3.3. **(Strong convergence)** *Let $u : [0,\infty) \to U$ be a trajectory satisfying (3.10′). Assume that the orbit $O(u) = \{u(t) : t \ge 0\}$ is precompact in X and has a cluster point $\varphi \in \mathcal{S}$ near which the gradient map $\mathcal{E}' : U \subset X \to Y \subset X'$ satisfies a gradient inequality, i.e., there exist some $\sigma > 0$ and $\phi \in \mathcal{G}$ such that*

$$\phi(\mathcal{E}(u)) \le \|\mathcal{E}'(u)\|_Y \quad \forall u \in U, \|u - \varphi\|_X < \sigma. \tag{$*$}$$

Then the trajectory u converges in X to φ in the sense that

$$\lim_{t\to\infty} \|u(t) - \varphi\|_X = 0 \quad \text{and} \quad I_\gamma(u) = \int_0^\infty \|\gamma\dot{u}(t)\|_{Y'}\,dt < \infty.$$

PROOF. We consider the bounded functions H and Φ used in the proof of Proposition 3.2. Note that the validity of the estimate in (3.16d) follows from the use of the gradient inequality (3.16c). Now for our present case where the gradient inequality $(*)$ is assumed, we see that (3.16c) holds true for all t in the set

$$W := \{t \geq 0 : \|u(t) - \sigma\|_X < \sigma\}.$$

Therefore, we have that

$$-\frac{d}{dt}\Phi(H(t)) \geq \|\gamma \dot{u}(t)\|_{Y'} \quad \text{a.e.} \quad t \in W.$$

An integration of this yields that

$$\int_W \|\gamma \dot{u}(t)\|_{Y'}\, dt \leq \Phi(H(0)) - \lim_{t\to\infty} \Phi(H(t)) < \infty.$$

Thus, we are in the position to apply Lemma 2.3, which yields the desired convergence $\|u(t) - \varphi\|_X \to 0$ as $t \to \infty$ as well as the integrability $\|\gamma \dot{u}(\cdot)\|_{Y'} \in L^1(\mathbb{R}_+)$. □

Now we turn to find some answers to the above problem for general trajectories satisfying the weaker growth condition (3.10a,b). First, we prepare some estimates.

LEMMA 3.4. *Let $u : [0,T] \to U$ be a trajectory satisfying (3.10a)–(3.10b).*
Assume further that the gradient map $\mathcal{E}' : U \to Y$ satisfies a gradient inequality of type $\mathcal{G}_k$ in the orbit $O(u)$; i.e., assume that the inequality

$$\phi(\mathcal{E}(u(t))) \leq \|\mathcal{E}'(u(t))\|_Y \quad \forall t \in [0,T] \tag{3.17}$$

holds true with a nonnegative function $\phi \in \mathcal{G}_k$ for some $k \geq 1$.
Then there holds

$$(1/k) \cdot \|\gamma \dot{u}(t)\|_{Y'} \leq -\frac{d}{dt}\Phi(H(t)) + \sqrt{F(t)} - G'(t) \tag{3.18a}$$

for a.e. $t \in [0,T]$, and

$$(1/k) \cdot \int_0^T \|\gamma \dot{u}(t)\|_{Y'}\, dt \leq \int_{H(T)}^{H(0)} \frac{ds}{\phi(s)} + G(0) + \int_0^T \sqrt{F(t)}\, dt, \tag{3.18b}$$

where both functions H and Φ are given by

$$H(t) := \mathcal{E}(u(t)) + F(t), \quad t \in [0,T] \tag{3.18c}$$

and

$$\Phi(x) := \int_0^x \frac{ds}{\phi(s)} \quad (x \in \mathbb{R}). \tag{3.18d}$$

PROOF. We have

$$-\dot{H}(t) = \langle -\mathcal{E}'(u(t)), \dot{u}(t)\rangle_{Y\times Y'} - F'(t).$$

It follows from the first inequality in (3.10a) that

$$-\dot{H}(t) \geq \|\mathcal{E}'(u(t))\|_Y \cdot \|\gamma \dot{u}(t)\|_{Y'} \tag{3.19}$$

and thus the absolutely continuous function H is nonincreasing. Therefore, an application of Lemma 2.1 yields that the composition $\Phi \circ H$ with the absolutely continuous function Φ given in (3.18d) is again absolutely continuous and

$$\frac{d}{dt}\Phi(H(t)) = \dot{\Phi}(H(t))\dot{H}(t) = \dot{H}(t)/\phi(H(t)) \quad a.e. \quad t \in [0,T].$$

To estimate $\phi(H(t))$, we use the following inequality for $\phi \in \mathcal{G}_k$:

$$\phi(x+y) \le k\phi(x) + |y|^{1/2} \quad \forall x, y \in \mathbb{R}.$$

It follows that

$$\phi(H(t)) = \phi(\mathcal{E}(u(t)) + F(t)) \le k\phi(\mathcal{E}(u(t))) + \sqrt{F(t)},$$

which, combined with (3.19), yields that

$$-\frac{d}{dt}\Phi(H(t)) \ge \frac{\|\gamma\dot{u}(t)\|_{Y'} \cdot \|\mathcal{E}'(u(t))\|_Y}{k\phi(\mathcal{E}(u(t))) + \sqrt{F(t)}}$$

and thus

$$-\frac{d}{dt}\Phi(H(t)) \ge \frac{\|\gamma\dot{u}(t)\|_{Y'} \cdot \|\mathcal{E}'(u(t))\|_Y}{k\|\mathcal{E}'(u(t))\|_Y + \sqrt{F(t)}}, \tag{3.20}$$

since $\phi(\mathcal{E}(u(t))) \le \|\mathcal{E}'(u(t))\|_Y$ for all $t \in [0,T]$ by the gradient inequality (3.17). Hence,

$$\begin{aligned} -k \cdot \frac{d}{dt}\Phi(H(t)) &\ge \frac{\|\gamma\dot{u}(t)\|_{Y'} \cdot (k\|\mathcal{E}'(u(t))\|_Y)}{k\|\mathcal{E}'(u(t))\|_Y + \sqrt{F(t)}} \\ &= \|\gamma\dot{u}(t)\|_{Y'} - \frac{\sqrt{F(t)} \cdot \|\gamma\dot{u}(t)\|_{Y'}}{k\|\mathcal{E}'(u(t))\|_Y + \sqrt{F(t)}}, \end{aligned}$$

which can be further estimated (using the fact $k \ge 1$) as

$$\|\gamma\dot{u}(t)\|_{Y'} \le (-k) \cdot \frac{d}{dt}\Phi(H(t)) + \sqrt{F(t)} \cdot \frac{\|\gamma\dot{u}(t)\|_{Y'}}{\|\mathcal{E}'(u(t))\|_Y + \sqrt{F(t)}}. \tag{3.21a}$$

To estimate the second term in the r.h.s. of (3.21a), we use the inequality in the second line of (3.10a), by noting the positivity $-G' \ge 0$, to find that

$$\frac{\|\gamma\dot{u}(t)\|_{Y'}}{\|\mathcal{E}'(u(t))\|_Y + \sqrt{F(t)}} \le \frac{\|\mathcal{E}'(u(t))\|_Y - G'(t)}{\|\mathcal{E}'(u(t))\|_Y + \sqrt{F(t)}} \le 1 - G'(t)/\sqrt{F(t)}. \tag{3.21b}$$

Taking together (3.21a) and (3.21b) and noting that $k \ge 1$, we have

$$(1/k) \cdot \|\gamma\dot{u}(t)\|_{Y'} \le -\frac{d}{dt}\Phi(H(t)) + \sqrt{F(t)} - G'(t) \quad \text{a.e.} \quad t \in [0,T],$$

which is the desired result in (3.18a). The estimate in (3.18b) follows from integrating (3.18a) over the interval $[0,T]$ and noting that $G(T) \ge 0$. □

As an immediate consequence of Lemma 3.4 we have the following weak convergence result.

THEOREM 3.5. **(Weak convergence)** *Let $u : [0, \infty) \to U$ be a trajectory satisfying a growth condition of the form (3.10a)–(3.10b).*

If the gradient map $\mathcal{E}' : U \to Y$ satisfies a gradient inequality of type $\mathcal{G}_k$ in the orbit $O(u)$, i.e., if there holds

$$\phi(\mathcal{E}(u(t))) \leq \|\mathcal{E}'(u(t))\|_Y \quad \forall t \geq 0$$

with a nonnegative function $\phi \in \mathcal{G}_k$ for some $k \geq 1$, then $\gamma u(t)$ converges as $t \to \infty$ with respect to the Y' norm in the sense that $I_\gamma(u) = \int_0^\infty \|\gamma \dot{u}(t)\|_{Y'}\, dt < \infty$.

The convergence in Y' is weaker than the convergence in X, since the Y' norm is smaller than or equal to the X norm; see (3.1c). However, under certain mild conditions (e.g., precompactness) on the trajectory, we have the convergence in X.

THEOREM 3.6. **(Strong convergence under the validity of gradient inequality of type $\mathcal{G}_k$)** *Let $u : [0, \infty) \to U$ be a trajectory satisfying (3.10a)–(3.10b). Assume the orbit $O(u) = \{u(t) : t \geq 0\}$ has a cluster point $\varphi \in \mathcal{S}$ near which the gradient map $\mathcal{E}' : U \subset X \to Y$ satisfies a gradient inequality of type $\mathcal{G}_k$, i.e., there exist $\sigma > 0$ and $\phi \in \mathcal{G}_k$ such that*

$$\phi(\mathcal{E}(v)) \leq \|\mathcal{E}'(v)\|_Y \quad \forall v \in U, \|v - \varphi\|_X < \sigma. \tag{3.22}$$

Furthermore, assume either (i) the orbit $O(u)$ is precompact in X or (ii) $Y = X'$ and the operator γ involved in (3.10a) is a homeomorphism on $Y' = X''$.

Then the trajectory u converges in X to φ in the sense that

$$\lim_{t\to\infty} \|u(t) - \varphi\|_X = 0 \quad \textit{and} \quad I_\gamma(u) = \int_0^\infty \|\gamma \dot{u}(t)\|_{Y'}\, dt < \infty.$$

PROOF. Let $W := \{t \geq 0 : \|u(t) - \varphi\|_X < \sigma\}$. Then the gradient inequality (3.22) ensures that Lemma 3.4 is applicable to any interval that is contained in the open set W. It follows (see (3.18a)) that

$$(1/k) \cdot \|\gamma \dot{u}(t)\|_{Y'} \leq -\frac{d}{dt}\Phi(H(t)) + \sqrt{F(t)} - G'(t) \quad \textit{a.e.} \quad t \in W.$$

Integrating this and using condition (3.10b), we find that

$$\int_W \|\gamma \dot{u}(t)\|_{Y'}\, dt < \infty. \tag{3.23a}$$

Thus, for case (i), we are in the position to apply Lemma 2.3 and obtain the desired convergence result $\|u(t) - \varphi\|_X \to 0$ as $t \to \infty$ as well as the integrability $\|\gamma \dot{u}(\cdot)\|_{Y'} \in L^1(\mathbb{R}_+)$.

For case (ii), where $Y = X'$ and γ is a homeomorphism on $Y' = X''$, we find from (3.23a) that

$$\int_W \|\gamma \dot{u}(t)\|_{X''}\, dt < \infty. \tag{3.23b}$$

Now a similar argument (by using (3.23b)) as that in the proof of Lemma 2.3 also yields the result that there exists some $s \geq 0$ such that $[s, \infty) \subset W$. This implies by (3.23b) again the integrability $\|\gamma \dot{u}(\cdot)\|_{X''} \in L^1(\mathbb{R}_+)$ as well as the convergence $\|u(t) - \varphi\|_X \to 0$ as $t \to \infty$. This completes the proof. $\square$

As seen in the previous chapter, analytic gradient maps satisfy gradient inequalities of LS-type, which is stronger than the $\mathcal{G}_k$ types needed in Theorem 3.6. Our following result implies that the growth condition (3.10b) of F, which describes the deviation of the trajectory u from a gradient trajectory generated by the gradient map $\mathcal{E}'$, can be improved a lot, once the gradient map satisfies a gradient inequality of weak LS-type.

We impose the following growth condition for both functions F and G involved in (3.10a):

$$(3.10b') \qquad \begin{cases} \lim_{t \to +\infty} (F(t) + G(t)) = 0 \quad \text{and} \\ \int_0^\infty \left(\frac{1}{\rho(t)} + \rho(t)|F'(t)| \right) dt < \infty \\ \text{with an increasing and positive function} \quad \rho. \end{cases}$$

The difference between (3.10b$'$) and (3.10b) is the growth condition on the nonincreasing absolutely continuous function F. Note that if F satisfies (3.10b), then for the increasing and positive function $\rho(t) := 1/\sqrt{F}$, we have that $1/\rho = \sqrt{F} \in L^1(\mathbb{R}_+)$ and $\rho|F'| = -F'/\sqrt{F} \in L^1(\mathbb{R}_+)$ and thus (3.10b$'$) is satisfied with $\rho = 1/\sqrt{F}$. This shows that (3.10b$'$) is far weaker than (3.10b).

THEOREM 3.7. **(Strong convergence under the validity of gradient inequality of weak LS-type)** *Let $u : [0, \infty) \to U$ be a trajectory satisfying (3.10a) and (3.10b$'$). Assume that the orbit $O(u) = \{u(t) : t \geq 0\}$ has a cluster point $\varphi \in \mathcal{S}$ near which the gradient map $\mathcal{E}' : U \subset X \to Y$ satisfies a gradient inequality of weak LS-type, i.e., there holds the gradient inequality*

$$(3.24a) \qquad \phi(|\mathcal{E}(v) - \mathcal{E}(\varphi)|) \leq \|\mathcal{E}'(v)\|_Y \quad \forall v \in U, \|v - \varphi\|_X < \sigma$$

with a constant $\sigma > 0$ and a (quasi-monotone) function $\phi \in \mathcal{G}$ for which there exists some constant $c > 0$ such that

$$(3.24b) \qquad \phi(y) \geq c \cdot \phi(x) \quad \forall\, 0 \leq x \leq y \leq 2x.$$

Furthermore, assume either (i) the orbit $O(u)$ is precompact in X or (ii) $Y = X'$ and the operator γ involved in (3.10a) is a homeomorphism on $Y' = X''$.

Then the trajectory u converges in X to φ in the sense that

$$\lim_{t \to \infty} \|u(t) - \varphi\|_X = 0 \quad \text{and} \quad I_\gamma(u) = \int_0^\infty \|\gamma \dot{u}(t)\|_{Y'}\, dt < \infty.$$

The proof of Theorem 3.7 is very similar to the proof for Theorem 3.6. However, the present manipulation is a little more delicate than before and we need one more elementary construction result.

LEMMA 3.8. *Let ρ, F be functions involved in condition (3.10b′) and assume furthermore that the C^1 function F is strictly decreasing. Let $\psi : \mathbb{R}_+ \to \mathbb{R}_+$ be the decreasing function such that $\psi(0) = 0$, $\psi(x) = \psi(F(0))$ for all $x \geq F(0)$, and*

$$\psi(F(t)) = 1/\rho(t) \quad \forall t \geq 0. \tag{3.25a}$$

Let $\phi \in \mathcal{G}$ be the quasi-monotone function given in Theorem 3.7. Then the function $\tilde{\phi}$ defined by

$$\tilde{\phi}(x) := \inf\{\phi(|x-y|) + \psi(|y|) : y \in \mathbb{R}\} \quad (x \in \mathbb{R}) \tag{3.25b}$$

is such that $\tilde{\phi} \in \mathcal{G}$ and

$$\tilde{\phi}(x+y) \leq \phi(|x|) + \psi(|y|) \quad \forall x, y \in \mathbb{R}. \tag{3.25c}$$

PROOF. The estimates in (3.25c) follow directly from the definition of $\tilde{\phi}$. It remains to prove $\tilde{\phi} \in \mathcal{G}$, i.e., $1/\tilde{\phi} \in L^1_{loc}(\mathbb{R})$. To this end, we note that

$$(*) \qquad \int_0^{F(0)} \frac{ds}{\psi(s)} = \int_0^\infty \frac{-dF(t)}{\psi(F(t))} = \int_0^\infty \rho(t)|F'(t)|\, dt < \infty.$$

The finiteness of the last integral is ensured by condition (3.10b′). Moreover, we have that

$$(**) \qquad \tilde{\phi}(x) \geq \min\{\psi(|x|/3), c \cdot \phi(2|x|/3)\} \quad \forall x \in \mathbb{R}$$

which is derived from the following observation: If y is such that $|y| > |x/3|$, then we have $\psi(|y|) \geq \psi(|x|/3)$ by the monotonicity of ψ; if y is such that $|y| \leq |x|/3$, then we have $2|x|/3 \leq |x-y| \leq 4|x|/3$ and thus $\phi(|x-y|) \geq c \cdot \phi(2|x|/3)$ by the quasi-monotonicity of ϕ. The local integrability of $1/\tilde{\phi}$ follows from $(*), (**)$ by noting that $\frac{1}{\phi} \in L^1_{loc}(\mathbb{R})$. □

Proof of Theorem 3.7. Without loss of generality, we can assume that the C^1 function F involved in both conditions (3.10a) and (3.10b′) is strictly decreasing. Hence, Lemma 3.8 is applicable and yields the function $\tilde{\phi} \in \mathcal{G}$ as constructed. As done in the proof of Lemma 3.4, with the nonincreasing and absolutely continuous function

$$H(t) := (\mathcal{E}(u(t)) - \mathcal{E}(\varphi)) + F(t) \quad (t \geq 0)$$

and the absolutely continuous function $\Phi(x) := \int_0^x (1/\tilde{\phi}(s))\, ds$ $(x \in \mathbb{R})$, we also have that

$$-\dot{H}(t) \geq \|\mathcal{E}'(u(t))\|_Y \cdot \|\gamma \dot{u}(t)\|_{Y'}$$

and

$$-\frac{d}{dt}\Phi(H(t)) = -\dot{H}(t)/\tilde{\phi}(H(t)) \quad \text{a.e.} \quad t \geq 0.$$

Consider the open and nonempty set $W := \{t \geq 0 : \|u(t) - \varphi\|_X < \sigma\}$. Let $t \in W$. By applying the gradient inequality (3.24a) to $v = u(t)$, we have that

$\phi(|\mathcal{E}(u(t)) - \mathcal{E}(\varphi)|) \le \|\mathcal{E}'(u(t))\|_Y$. It follows that

$$\begin{aligned}\tilde{\phi}(H(t)) &= \tilde{\phi}((\mathcal{E}(u(t)) - \mathcal{E}(\varphi)) + F(t)) \\ &\le \phi(|\mathcal{E}(u(t)) - \mathcal{E}(\varphi)|) + \psi(F(t)) \\ &\le \|\mathcal{E}'(u(t))\|_Y + \psi(F(t)).\end{aligned}$$

For the first inequality, we have used the property $\tilde{\phi}(x+y) \le \phi(|x|) + \psi(|y|)$ ($x, y \in \mathbb{R}$) given by (3.25c). Taking these estimates together, we obtain that

$$-\frac{d}{dt}\Phi(H(t)) \ge \frac{\|\mathcal{E}'(u(t))\|_Y \cdot \|\gamma\dot{u}(t)\|_{Y'}}{\|\mathcal{E}'(u(t))\|_Y + \psi(F(t))} \quad \text{a.e.} \quad t \in W. \tag{3.26a}$$

Note the similarity of (3.26a) and (3.20). Hence, a completely similar manipulation under the use of the estimate $\|\gamma\dot{u}(t)\|_{Y'} \le \|\mathcal{E}'(u(t))\|_Y - G'(t)$ in (3.10a) yields that

$$\|\gamma\dot{u}(t)\|_{Y'} \le -\frac{d}{dt}\Phi(H(t)) + \psi(F(t)) - G'(t) \quad a.e. \quad t \in W.$$

By (3.25a) and (3.10b′) we have that $\psi \circ F = 1/\rho \in L^1(\mathbb{R}_+)$. Thus, by taking integration over W in the above estimate and noting that all integrands are nonnegative, we find that

$$\int_W \|\gamma\dot{u}(t)\|_{Y'}\, dt \le \Phi(H(0)) + \int_0^\infty \frac{dt}{\rho(t)} + G(0) < \infty. \tag{3.26b}$$

As done in the proof of Theorem 3.6, (3.26b) has been sufficient to ensure the desired convergence $\|u(t) - \varphi\|_X \to 0$ as $t \to \infty$ as well as the finiteness $I_\gamma(u) < \infty$. □

4. Convergence in Hilbert spaces

In this section we will use Theorem 3.7 and both Lemmas 2.2 and 2.3 to derive convergence results for trajectories defined on Hilbert spaces. The Hilbert space setting arises from the most concrete application situations.

We let H be a real Hilbert space and let X be a subspace of H that becomes a Banach space under its own norm $\|\cdot\|_X$. Assume $\mathcal{E} : X \to \mathbb{R}$ to be a C^1 functional that has the property that $\mathcal{E}'(X) \subset H$ and such that the map

$$\mathcal{E}' : X \to H \quad \text{is continuous.}$$

We choose

$$Y := H \cong Y'.$$

Then condition (3.10a) corresponding to the choice $Y = H \cong Y'$ reads as follows:

$$\begin{cases}\langle -\mathcal{E}'(u(t)), \dot{u}(t)\rangle_{H\times H} \ge \|\mathcal{E}'(u(t))\|_H \cdot \|\gamma\dot{u}(t)\|_H + F'(t), \\ \|\mathcal{E}'(u(t))\|_H \ge \|\gamma\dot{u}(t)\|_H + G'(t)\end{cases} \tag{3.27a}$$

for all $t \ge 0$, where $\gamma : H \to H$ is an *injective* and continuous map (which could be *nonlinear*), and both $F, G : \mathbb{R}_+ \to \mathbb{R}_+$ are nonincreasing absolutely continuous functions. We recall that the growth conditions (3.10b) and (3.10b′) for both F and G are respectively

$$\text{(3.27b)} \qquad \lim_{t\to+\infty}(F(t)+G(t)) = 0 \quad \text{and} \quad \int_0^\infty \sqrt{F(t)}\,dt < \infty,$$

and

$$\text{(3.27b}'\text{)} \qquad \lim_{t\to+\infty}(F(t)+G(t)) = 0 \quad \text{and} \quad \int_0^\infty \left(\frac{1}{\rho(t)} - \rho(t)F'(t)\right) dt < \infty$$

with some increasing positive function ρ.

Let $\varphi \in X$. Recall that the gradient map $\mathcal{E}' : X \to H$ is said to satisfy a gradient inequality of type $\mathcal{G}_k$ (with a constant $k \geq 1$) near φ if there exists some constant $\sigma > 0$ such that

$$(*) \qquad \phi(\mathcal{E}(u)) \leq \|\mathcal{E}'(u)\|_H \quad \forall u \in X, \|u-\varphi\|_X < \sigma$$

with a function ϕ satisfying $0 \leq \frac{1}{\phi} \in L^1_{loc}(\mathbb{R}_+)$ and that

$$\phi(x+y) \leq k\phi(x) + \sqrt{|y|} \quad \forall x, y \in \mathbb{R}.$$

The gradient inequality $(*)$ is said to be of LS-type if there exist constants $c, \sigma > 0$ and $\theta \in [1/2, 1)$ such that

$$(**) \qquad c \cdot |\mathcal{E}(u) - \mathcal{E}(\varphi)|^\theta \leq \|\mathcal{E}'(u)\|_H \quad \forall u \in X, \|u-\varphi\|_X < \sigma.$$

Moreover, the gradient inequality $(*)$ is said to be of weak LS-type if there holds

$$(***) \qquad \phi(|\mathcal{E}(u) - \mathcal{E}(\varphi)|) \leq \|\mathcal{E}'(u)\|_H \quad \forall u \in X, \|u-\varphi\|_X < \sigma$$

with a constant $\sigma > 0$ and a (quasi-monotone) function $\phi \in \mathcal{G}$ for which there exists some constant $c > 0$ such that

$$(****) \qquad \phi(y) \geq c \cdot \phi(x) \quad \forall 0 \leq x \leq y \leq 2x.$$

We have seen in Chapter 2 that a lot of analytic and non-analytic gradient maps enjoy gradient inequalities of LS-type.

4A: Convergence in first-order systems. We first treat the simplest case where the gradient system

$$(*) \qquad \dot{u} = -\mathcal{E}'(u)$$

is under consideration. We want to prove Theorem 1.1 of Chapter 1, which we restate again for the sake of completeness.

PROPOSITION 4.1. *Assume that $u : \mathbb{R}_+ \to X$ is a classical solution of the gradient system $(*)$ such that*

$$\sup\{|\mathcal{E}(u(t))| : t \geq 0\} < \infty.$$

Then the following assertions **(i)–(iii)** *are equivalent:*

(i) *There exists some $\phi \in \mathcal{G}$ such that*

$$\phi(\mathcal{E}(v)) \leq \|\mathcal{E}'(v)\|_H \quad \forall v \in O(u).$$

(ii) *There holds*

$$I(u) = \int_0^\infty \|\dot{u}(t)\|_H \, dt < \infty.$$

(iii) *There exists some function* $\tilde{\phi} \in \mathcal{G}$ *such that*

$$\int_0^\infty \tilde{\phi}(\mathcal{E}(u(t))) \, dt < \infty.$$

PROOF. The equivalence **(i)**$\Longleftrightarrow$**(ii)** follows directly from Proposition 3.1 combined with Proposition 3.2 under the choices $Y = H \cong Y'$ and γ=Identity operator on H.

We will establish the equivalence of **(iii)** to both **(i)** and **(ii)** by showing the following implications: **(i)**$\Longrightarrow$**(iii)**$\Longrightarrow$**(ii)**.

To this end, we assume **(i)**. Then we have that

$$\|\mathcal{E}'(u(\cdot))\|_H = \|\dot{u}(\cdot)\|_H \in L^1(\mathbb{R}_+)$$

by the implication **(i)**$\Longrightarrow$**(ii)**. Using the gradient inequality in **(i)**, we have that

$$\phi(\mathcal{E}(u(t))) \leq \|\mathcal{E}'(u(t))\|_H$$

for all $t \geq 0$ and thus $\phi(\mathcal{E}(u(\cdot))) \in L^1(\mathbb{R}_+)$, giving **(iii)**.

Now we assume **(iii)** and consider the C^1 function

$$h(t) := \mathcal{E}(u(t)) \quad (t \geq 0).$$

We have that

$$h'(t) = \langle \mathcal{E}'(u(t)), \dot{u}(t) \rangle_{H \times H} = -\|\dot{u}(t)\|_H^2.$$

Therefore, the C^1 function h is bounded and nonincreasing. Moreover, by **(iii)** we have that $\tilde{\phi}(h(\cdot)) = \tilde{\phi}(\mathcal{E}(u(\cdot))) \in L^1(\mathbb{R}_+)$ with some function $\tilde{\phi} \in \mathcal{G}$. Therefore, we are in the position to apply Lemma 2.2 under the choice $\Lambda := \mathbb{R}_+$ and $g(\cdot) := \tilde{\phi}(h(\cdot)) \in L^1(\mathbb{R}_+)$. It yields that

$$\|\dot{u}(\cdot)\|_H = \sqrt{|h'(\cdot)|} \in L^1(\mathbb{R}_+),$$

giving **(ii)**. This completes the proof. □

Next we consider the following more general first order gradient-like system in the Hilbert space H :

$$\begin{cases} \dot{u} = -\mathcal{N}(u) + g(t,u), & t \geq 0, \\ u \equiv u(t) \in H, \dot{u}(t) \in H, \end{cases} \tag{3.28a}$$

where $\mathcal{N} : X \to H$ is a continuous map for which there exist a **strictly positive** continuous function $\eta : X \to (0,1)$ and an **injective** linear and bounded operator $\tilde{J} : H \to H$ such that

$$\begin{cases} \langle \mathcal{E}'(u), \mathcal{N}(u) \rangle_{H \times H} \geq \eta(u) \cdot \|\mathcal{E}'(u)\|_H^2 \\ \|\tilde{J}\mathcal{N}(u)\|_H \leq (1/\eta(u)) \cdot \|\mathcal{E}'(u)\|_H \end{cases} \quad \forall u \in X. \tag{3.28b}$$

Moreover, the time-dependent perturbation $g : \mathbb{R}_+ \times X \to H$ is a continuous map.

Roughly speaking, condition (3.28b) says that $\mathcal{N}$ is a gradient-like map associated to the gradient map $\mathcal{E}' : X \to H$. Moreover, note that condition (3.28b) is satisfied for the special choice

$$\mathcal{N} = J\mathcal{E}'$$

where

$$J : D(J) \subset H \to H$$

is a closed, surjective and coercive **linear** operator. (For this case, the choice $\tilde{J} := J^{-1}$ does the job.)

As seen in the previous section, if the gradient map $\mathcal{E}'$ has the property that it satisfies a gradient inequality of LS-type near every critical point of $\mathcal{E}$, then every solution to (3.28a) whose orbit is precompact in X will converge to some critical point of $\mathcal{E}$ with respect to the X norm topology whenever the perturbation g is small in some suitable sense.

The smallness conditions that we have to impose on the perturbations $g(t,u)$ are the following two types:

$$\text{(3.28c)} \qquad \begin{cases} \|g(s,u)\|_H \leq g(s) \quad \forall s \geq 0, u \in X; \\ \displaystyle\int_0^\infty \left(\int_t^\infty g(s)^2\, ds \right)^{1/2} dt < \infty, \end{cases}$$

and

$$\text{(3.28c}'\text{)} \qquad \begin{cases} \|g(s,u)\|_H \leq g(s) \quad \forall s \geq 0, u \in X; \\ \displaystyle\int_0^\infty \rho(t) g(t)^2\, dt + \int_0^\infty \frac{1}{\rho(t)}\, dt < \infty \\ \text{with some increasing positive function} \quad \rho. \end{cases}$$

We take a look at these growth conditions. As observed in Huang-Takáč [**70**], we have the following implication

$$\text{(3.28c)} \Longrightarrow \int_0^\infty (1+t)^2 g(t)^2\, dt < \infty.$$

From this we see that (3.28c) is stronger than (3.28c'). A specially interesting case of (3.28c') is the following

$$\text{(3.28c}''\text{)} \qquad \begin{cases} \|g(t,u)\|_H \leq g(t) \quad \forall t \geq 0, u \in X; \\ \displaystyle\int_0^\infty t|\log t|^\tau g(t)^2\, dt < \infty \quad \text{with some} \quad \tau > 1 \end{cases}$$

corresponding to the choice $\rho(t) = t(1 + \log(1+t))^\tau$ $(t \geq 0)$ with $\tau > 1$.

We have the following implication of (3.28c$'$):

$$\text{(3.28c}'''\text{)} \qquad g \in L^2(\mathbb{R}_+) \quad \text{and} \quad g \in L^1(\mathbb{R}_+).$$

That $g \in L^2(\mathbb{R}_+)$ is clear. To see the second one, we note that

$$2g(t) \leq \rho(t)g(t)^2 + \frac{1}{\rho(t)}.$$

Therefore, $g \in L^1$ follows from condition (3.28c′).

It is also worth noting the following fact: Let $0 \leq g \in L^2(\mathbb{R}_+)$ be a *decreasing* function. Then the condition that g satisfies (3.28c′), i.e., the existence of some increasing positive function ρ such that

$$(*) \qquad \int_0^\infty \rho(t)g(t)^2\,dt + \int_0^\infty \frac{1}{\rho(t)}\,dt < \infty$$

is equivalent to the very simple condition

$$(**) \qquad g \in L^1(\mathbb{R}_+).$$

In fact, by (3.28c‴), we have the implication $(*) \Longrightarrow (**)$. On the other hand, if $g \in L^1$, then the increasing function $\rho(t) := 1/(g(t) + e^{-t})$ is such that $(**)$ holds true.

Note that $(**)$ is merely the necessary condition for the convergence of solutions to the one-dimensional equation $\dot{u}(t) = g(t)$. Hence, the growth condition (3.28c′) on the perturbations also serves to some extent as a necessary condition for the following convergence result.

THEOREM 4.2. *Under the above situations we let $u : [0, \infty) \to X$ be a solution of (3.28a,b) such that its orbit $O(u) = \{u(t) : t \geq 0\}$ is precompact in X. Assume, in addition, that one of the following is satisfied:*

(i) The time-dependent perturbation $g(t, u)$ in (3.28a) satisfies the growth condition (3.28c) and there exists some cluster point $\varphi \in X$ of the orbit $O(u)$ near which the gradient map $\mathcal{E}' : X \to H$ satisfies a gradient inequality of type $\mathcal{G}_k$ with some $k \geq 1$.

(ii) The time-dependent perturbation $g(t, u)$ in (3.28a) satisfies the growth condition (3.28c′) and there exists some cluster point $\varphi \in X$ of the orbit $O(u)$ near which the gradient map $\mathcal{E}' : X \to H$ satisfies a gradient inequality of weak LS-type.

Then there holds

$$\lim_{t\to\infty} \|u(t) - \varphi\|_X = 0, \quad \mathcal{E}'(\varphi) = 0.$$

PROOF. We will apply Theorem 3.6 to case (i) and Theorem 3.7 to case (ii), under the choice of the space $Y = H \cong Y'$. To do so, we need to verify that the solution u is a trajectory satisfying condition (3.27a) with the choice $\gamma = const \cdot \tilde{J}$ and some suitable functions F and G that satisfy condition (3.27b) for case (i) and condition (3.27b′) for case (ii).

Below we will denote the scalar product on H simply by $\langle \cdot, \cdot \rangle$, and the norm on H simply by $\| \cdot \|$.

Let

$$(3.29a) \qquad c_1 := \inf\{\eta(v) : v \in O(u)\} \in [0, 1).$$

Since $O(u)$ is precompact in X, the closure $K := O(u)^{\mathrm{cl}}$ is compact in X. Therefore, the continuity combined with the strict positivity of the function η implies $c_1 \geq \min\{\eta(v) : v \in K\} > 0$. Hence, by condition (3.28b) we have that

$$\text{(3.29b)} \qquad \begin{cases} \langle \mathcal{E}'(u(t)), \mathcal{N}(u(t))\rangle \geq c_1 \cdot \|\mathcal{E}'(u(t))\|^2 \\ \|\tilde{J}\mathcal{N}(u(t))\| \leq (1/c_1) \cdot \|\mathcal{E}'(u(t))\| \end{cases} \qquad \forall t \geq 0.$$

Using the equation $\dot{u} = -\mathcal{N}(u) + g(t,u)$, we find that

$$\langle -\mathcal{E}'(u), \dot{u}\rangle = \langle \mathcal{E}'(u), \mathcal{N}(u)\rangle - \langle \mathcal{E}'(u), g(t,u)\rangle.$$

The estimate in the first line of (3.29b) combined with the estimate

$$|\langle \mathcal{E}'(u), g(t,u)\rangle| \leq \|\mathcal{E}'(u)\| \cdot g(t) \leq \frac{1}{2}(c_1\|\mathcal{E}'(u)\|^2 + g(t)^2/c_1)$$

yields that

$$\text{(3.30a)} \qquad \langle -\mathcal{E}'(u(t)), \dot{u}(t)\rangle \geq (c_1/2) \cdot \|\mathcal{E}'(u(t))\|^2 - g(t)^2/(2c_1).$$

On the other hand, we have that

$$\|\tilde{J}\dot{u}(t)\| = \|\tilde{J}\mathcal{N}(u(t)) - \tilde{J}g(t,u(t))\| \leq \|\tilde{J}\mathcal{N}(u(t))\| + \|\tilde{J}\| \cdot g(t),$$

which yields by the estimate in the second line of (3.29b) that

$$\text{(3.30b)} \qquad \|\mathcal{E}'(u(t))\| \geq c_1 \cdot \|\tilde{J}\dot{u}(t)\| - (c_1\|\tilde{J}\|) \cdot g(t).$$

We use (3.30b) to estimate the right-hand side of (3.30a). It yields that

$$\text{(3.30c)} \qquad \langle -\mathcal{E}'(u(t)), \dot{u}(t)\rangle \geq (c_1/2)^2 \cdot \|\mathcal{E}'(u(t))\| \cdot \|\tilde{J}\dot{u}(t)\| - c_2 \cdot g(t)^2$$

with $c_2 := (1/c_1 + (c_1\|\tilde{J}\|)^2)$. Comparing (3.30b,c) with (3.27a), we see that (3.27a) is satisfied with the choices $\gamma := \min\{c_1, c_1^2/4\} \cdot \tilde{J}$ and

$$F(t) := c_2 \cdot \int_t^\infty g(s)^2\, ds, \quad G(t) := (c_1\|\tilde{J}\|) \cdot \int_t^\infty g(s)\, ds.$$

Both integrals exist because $g \in L^1 \cap L^2(\mathbb{R}_+)$ (see (3.28c″)). This implies that both F and G are nonincreasing absolutely continuous functions such that $F(t) + G(t) \to 0$ as $t \to \infty$. We have

$$0 \leq -F'(t) = g(t)^2 \quad a.e. \quad t \geq 0.$$

Thus F satisfies condition (3.27b) for case (i) where (3.28c) is satisfied, and F satisfies condition (3.27b') for case (ii) where (3.28c') is satisfied, by taking the same function ρ involved in both (3.27b′) and (3.28c′).

So far we are in the position to apply Theorem 3.6 for case (i) and Theorem 3.7 for case (ii). It yields the convergence $u(t) \to \varphi$ in X as $t \to \infty$ as well as the convergence $\|\tilde{J}\dot{u}(\cdot)\| \in L^1(\mathbb{R}_+)$. The latter implies that $\|\tilde{J}\mathcal{N}(u(\cdot))\| \in L^1(\mathbb{R}_+)$. Since the map $\tilde{J} \circ \mathcal{N} : X \to H$ is continuous, we conclude from above that $\tilde{J}\mathcal{N}(\varphi) = 0$ and thus $\mathcal{N}(\varphi) = 0$ by the injectivity of $\tilde{J}$. This finally yields that $\mathcal{E}'(\varphi) = 0$, since there holds $\|\mathcal{N}(u(t))\| \geq$

$c_1\|\mathcal{E}'(u(t))\|$ for all $t \geq 0$ by the estimate in the first line of (3.29b) and the result follows by letting $t \to \infty$. We have completed the proof. □

REMARK 4.3. As seen in Example 3.5 of Chapter 2, there are non-analytic functions V on $\mathbb{R}^2$ given by (2.0) whose gradient satisfies a gradient inequality of weak LS-type in every compact subset. Hence, Theorem 4.2 (ii) applies to the corresponding perturbed planar system

$$\dot{u} = -\nabla V(u) + p(t). \tag{3.31a}$$

It yields that a bounded solution of (3.31a) will converge to some equilibrium, if the perturbation p is such that

$$\int_0^\infty t|\log t|^\tau |p(t)|^2\,dt < \infty \tag{3.31b}$$

with some $\tau > 1$, cf. (3.28c′). In Chapter 5 we will give numerical results which strongly suggest that the exponent τ in the growth condition (3.31b) cannot be reduced down to the bound 1. Hence, a growth condition of the form (3.28c′) could be considered an optimal one.

4B: Convergence in second-order systems. Next we consider the following second-order gradient-like systems

$$\begin{cases} \ddot{u} + \kappa B(t,u,\dot{u}) = \pm\mathcal{E}'(u) + g(t,u,\dot{u}), \quad t \geq 0, \\ u \equiv u(t) \in H, \dot{u} \equiv \dot{u}(t) \in H, \ddot{u} \equiv \ddot{u}(t) \in H, \end{cases} \tag{3.32a}$$

for which the following conditions (i)–(iv) are assumed:

(i) The gradient map $\mathcal{E}' : X \to H$ is of C^1 class.

(ii) The constant κ is either equal to 1 or -1.

(iii) $B : \mathbb{R}_+ \times X \times X \to H$ is a continuous map that is *coercive* in the sense that there exists a constant $c > 0$ such that

$$\langle B(t,u,v), v\rangle_{H\times H} \geq c \cdot \|v\|_H^2 \quad \forall (t,u,v) \in \mathbb{R}_+ \times X \times X. \tag{3.32b}$$

(iv) The perturbation $g : \mathbb{R}_+ \times X \times H \to H$ is a continuous map and satisfies the following smallness condition:

$$\begin{cases} \sup_{u\in X, v\in H} \|g(s,u,v)\|_H \leq g(s) \quad \forall s \geq 0; \\ \displaystyle\int_0^\infty \left(\int_t^\infty g(s)^2\,ds\right)^{1/2} dt < \infty. \end{cases} \tag{3.32c}$$

From the coercivity of B we see that the choice $\kappa = 1$ corresponds to the case that $\kappa B(t,u,\dot{u})$ is a *damping*, while the choice $\kappa = -1$ reflects that the term $\kappa B(t,u,\dot{u})$ is an *excitation*.

For the case when $\mathcal{E}'(u) = -\Delta u + f(x,u)$ is a semilinear elliptic operator, the equation (3.32a) has two types corresponding to the choice of the sign before $\mathcal{E}'(u)$. With the positive sign, the equation (3.32a) becomes the semilinear equation of elliptic type

$$\ddot{u}(t) + \Delta u + \kappa B(t,u,\dot{u}) = f(x,u) + g(t,u,\dot{u}).$$

Hence, we will call (3.32a) with the positive sign gradient-like systems of *elliptic type*. With the negative sign, the equation (3.32a) becomes the semilinear wave equation

$$\ddot{u}(t) + \kappa B(t, u, \dot{u}) = \Delta u - f(x, u) + g(t, u, \dot{u}).$$

Hence, we will call (3.32a) with the negative sign gradient-like systems of *wave type*.

For gradient-like systems of elliptic type, we have the following convergence result that improves results previously obtained by Simon [**107**], Jendoubi [**71, 72**], and Huang and Takáč [**70**, Theorem 3.2]. Our tools for the proof are Lemmas 2.2 and 2.3 of this chapter.

THEOREM 4.4. **(Convergence in gradient-like systems of elliptic type)** *Under the above situations (where conditions (i)–(iv) are satisfied), we let $u : [0, \infty) \to X$ be a solution of (3.32a) with the* **positive sign**. *Assume that*

$$\{\dot{u}(t) : t \geq 0\} \subset X, \quad \sup\{\|\dot{u}(t)\|_H : t \geq 0\} < \infty,$$

and there exists a constant $c_1 > 0$ such that

$$\langle \mathcal{E}''(u(t))\dot{u}(t), \dot{u}(t)\rangle_{H\times H} \geq -c_1 \cdot \|\dot{u}(t)\|_H^2 \quad \forall t \geq 0 \tag{3.33a}$$

and

$$\|B(t, u(t), \dot{u}(t))\|_H \leq c_1 \cdot \|\dot{u}(t)\|_H \quad \forall t \geq 0. \tag{3.33b}$$

Assume furthermore that the orbit $O(u) = \{u(t) : t \geq 0\}$ is precompact in X and there exists some cluster point $\varphi \in X$ of the orbit $O(u)$ near which the gradient map $\mathcal{E}' : X \to H$ satisfies a gradient inequality of type $\mathcal{G}_k$ with some $k \geq 1$, i.e., there exist a constant $\sigma > 0$ and a function $\phi \in \mathcal{G}_k$ such that

$$\phi(\mathcal{E}(w)) \leq \|\mathcal{E}'(w)\|_H \quad \forall w \in X, \|w - \varphi\|_X < \sigma. \tag{3.33c}$$

Then there holds

$$\lim_{t\to\infty} (\|\dot{u}(t)\|_H + \|u(t) - \varphi\|_X) = 0, \quad \mathcal{E}'(\varphi) = 0.$$

PROOF. The proof is divided into three steps.

Step 1. Let α be a positive constant whose smallness will be determined later. Consider the function

$$h(t) := -\frac{\kappa}{2}\|\dot{u}(t)\|_H^2 + \kappa\mathcal{E}(u(t)) + \alpha\langle \mathcal{E}'(u(t)), \dot{u}(t)\rangle_{H\times H} \quad (t \geq 0). \tag{3.34}$$

We have, with $u \equiv u(t) \in X$ and $v \equiv \dot{u}(t) \in X$, that

$$h'(t) = \langle -\kappa v + \alpha\mathcal{E}'(u), \ddot{u}\rangle + \kappa\langle \mathcal{E}'(u), v\rangle + \alpha\langle \mathcal{E}''(u)v, v\rangle.$$

Here and below, we use $\langle \cdot, \cdot\rangle$ and $\|\cdot\|$ to denote the scalar product and the norm on the Hilbert space H, respectively.

Under the substitution $\ddot{u} = \mathcal{E}'(u) + g(t, u, v) - \kappa B(t, u, v)$ and using the equality $\kappa^2 = 1$, we obtain that

$$h'(t) = \langle v, B\rangle + \alpha\|\mathcal{E}'(u)\|^2 + \alpha\langle \mathcal{E}''(u)v, v\rangle + \tilde{h}(t),$$

where $B \equiv B(t,u,v)$ and

$$\tilde{h}(t) := -\alpha\kappa\langle \mathcal{E}'(u), B\rangle - \langle \kappa v - \alpha\mathcal{E}'(u), g(t,u,v)\rangle.$$

Using the Schwarz inequality combined with the elementary inequality $2|ab| \le a^2 + b^2$ $(a, b \in \mathbb{R})$, we have that

$$\begin{aligned} |\kappa\langle \mathcal{E}'(u), B\rangle| &\le \|\mathcal{E}'(u)\| \cdot \|B\| \\ &\le \|\mathcal{E}'(u)\| \cdot (c_1\|v\|) \quad \text{(by (3.33b))} \\ &\le \|\mathcal{E}'(u)\|^2/4 + c_1^2\|v\|^2 \end{aligned}$$

and

$$\begin{aligned} &|\langle \kappa v - \alpha\mathcal{E}'(u), g(t,u,v)\rangle| \le (\|v\| + \alpha\|\mathcal{E}'(u)\|) \cdot g(t) \quad \text{(by (3.32c))} \\ &\le c\|v\|^2/2 + \alpha\|\mathcal{E}'(u)\|^2/4 + (\alpha + 1/c^2)g(t)^2. \end{aligned}$$

Here $c > 0$ is the constant involved in (3.32b). It follows that

$$|\tilde{h}(t)| \le \alpha\|\mathcal{E}'(u)\|^2/2 + (c/2 + \alpha c_1^2)\|v\|^2 + (\alpha + 1/c^2)g(t)^2.$$

Therefore, by using the inequality $\langle v, B\rangle \ge c\|v\|^2$ in (3.32b) as well as the inequality in (3.33a), we find from above that

$$h'(t) \ge (c/2 - \alpha c_1 - \alpha c_1^2) \cdot \|v\|^2 + (\alpha/2)\|\mathcal{E}'(u)\|^2 - (\alpha + 1/c^2)g(t)^2.$$

Thus, if

(3.35a) $$0 < \alpha < c/(2c_1 + 2c_1^2),$$

then we have (by recovering $u = u(t), v = \dot{u}(t)$) that

$$h'(t) \ge c_2 \cdot (\|\dot{u}(t)\| + \|\mathcal{E}'(u(t))\|)^2 - c_3 g(t)^2 \quad \forall t \ge 0$$

with $c_2 := \frac{1}{4}\min\{\alpha, c - 2\alpha(c_1 + c_1^2)\} > 0$ and $c_3 := \alpha + 1/c^2$. This implies that the new function

(3.35b) $$H(t) := h(t) + c_3\int_t^\infty g(s)^2\,ds \quad (t \ge 0)$$

is nondecreasing and absolutely continuous, such that

(3.35c) $$H'(t) \ge c_2 \cdot (\|\dot{u}(t)\| + \|\mathcal{E}'(u(t))\|)^2 \quad \forall t \ge 0.$$

Step 2. By assumption, the norms $\|\dot{u}(t)\|$ $(t \ge 0)$ are uniformly bounded. Moreover, the precompactness of the orbit $O(u)$ in X implies the uniform boundedness of the function $t \mapsto \mathcal{E}(u(t))$ as well as the norms $\|\mathcal{E}'(u(t))\|$ $(t \ge 0)$. Hence, the function $H(t)$ is uniformly bounded and thus, by noting $g \in L^2(\mathbb{R}_+)$, we conclude from (3.35c) that $\|\mathcal{E}'(u(\cdot))\| + \|\dot{u}(\cdot)\| \in L^2(\mathbb{R}_+)$. This implies by (3.33b) that $\|B(\cdot, u(\cdot), \dot{u}(\cdot))\| \in L^2(\mathbb{R}_+)$ and thus, by noting $\ddot{u} = \mathcal{E}'(u) - \kappa B(t,u,\dot{u}) + g(t,u,\dot{u})$, we find that $\|\ddot{u}(\cdot)\| \in L^2(\mathbb{R}_+)$. It follows that

$$\frac{d}{dt}\|\dot{u}(t)\|^2 = 2\langle \dot{u}(t), \ddot{u}(t)\rangle \in L^1(\mathbb{R}_+).$$

Therefore, we have the convergence

$$\lim_{t\to\infty}\|\dot{u}(t)\|^2 = \lim_{t\to\infty}\int_t^\infty 2\langle \dot{u}(s), \ddot{u}(s)\rangle\,ds = 0.$$

Step 3. Let φ be the cluster point of $O(u)$ given in the theorem. We want to show the convergence $u(t) \longrightarrow \varphi$ as $t \longrightarrow \infty$ in X. For this purpose, we use the gradient inequality (3.33c) and the assumption that $\phi \in \mathcal{G}_k$ with some $k \geq 1$ to find that there holds

$$\phi_1(\kappa\mathcal{E}(u)) \leq \|\mathcal{E}'(u)\| \quad \forall u \in X, \|u - \varphi\|_X < \sigma, \tag{3.36a}$$

and

$$\phi_1(x + y) \leq k\phi_1(x) + \sqrt{|y|} \quad (x, y \in \mathbb{R}) \tag{3.36b}$$

with the new function $\phi_1 \in \mathcal{G}_k$ given by $\phi_1(s) := \phi(\kappa s)$ ($\forall s \in \mathbb{R}$). (Remember that $\kappa^2 = 1$.)

Consider the function H given by (3.35b) (see also (3.34) for the definition of h). We have that

$$H(t) = \kappa\mathcal{E}(u(t)) + (\delta(t) + g_1(t)),$$

where

$$\delta(t) := -\kappa\|\dot{u}(t)\|^2/2 + \alpha\langle\mathcal{E}'(u(t)), \dot{u}(t)\rangle, \quad g_1(t) := c_3 \int_t^\infty g(s)^2\, ds.$$

We have that

$$|\delta(t)| \leq \|\dot{u}\|^2/2 + \alpha\|\mathcal{E}'(u)\| \cdot \|\dot{u}\| \leq \|\dot{u}\|^2 + (\alpha\|\mathcal{E}'(u)\|)^2.$$

Therefore, by using (3.36b) and the elementary inequality $\sqrt{|a + b|} \leq \sqrt{|a|} + \sqrt{|b|}$ ($a, b \in \mathbb{R}$) twice, we obtain that

$$\phi_1(H(t)) \leq k\phi_1(\kappa\mathcal{E}(u(t))) + \alpha\|\mathcal{E}'(u(t))\| + \|\dot{u}(t)\| + \sqrt{g_1(t)}.$$

Let

$$W := \{t \geq 0 : \|u(t) - \varphi\|_X < \sigma\}.$$

Then, for $t \in W$ we have by the gradient inequality (3.36a) that $\phi_1(\kappa\mathcal{E}(u(t))) \leq \|\mathcal{E}'(u(t))\|$, and thus

$$\phi_1(H(t)) \leq (k + \alpha)\|\mathcal{E}'(u(t))\| + \|\dot{u}(t)\| + \sqrt{g_1(t)}.$$

It follows, by using (3.35c), that

$$\phi_1(H(t)) \leq ((k + \alpha)/\sqrt{c_2})\sqrt{H'(t)} + \sqrt{g_1(t)} \quad \forall t \in W.$$

Consequently, with the function $\phi_2 := \sqrt{c_2}/(2k + 2\alpha)\phi_1 \in \mathcal{G}$, we have that

$$\phi_2(H(t)) \leq \sqrt{H'(t)} + \sqrt{g_1(t)} \quad \forall t \in W.$$

By (3.32c), $\sqrt{g_1} \in L^1(\mathbb{R}_+)$. Therefore, the function H satisfies the inequalities (3.8) of Lemma 2.2, and thus $\sqrt{H'(\cdot)} \in L^1(W)$ by Lemma 2.2 of this chapter. This implies by (3.35c) that

$$\int_W \|\dot{u}(t)\|\, dt < \infty.$$

Therefore, we are in the position to apply Lemma 2.3 of this chapter. It yields the convergence $\|u(t) - \varphi\|_X \longrightarrow 0$ as $t \longrightarrow \infty$. Finally, the assertion $\mathcal{E}'(\varphi) = 0$ follows from the convergence $\|\mathcal{E}'(u(t)) - \mathcal{E}'(\varphi)\| \longrightarrow 0$ as $t \longrightarrow \infty$

combined with the integrability $\|\mathcal{E}'(u(\cdot))\| \in L^2(\mathbb{R}_+)$, which is ensured by (3.35c). This completes the proof. □

For gradient-like systems of wave type we have the following result that unifies and extends many results previously obtained by Chill-Jendoubi [**21**, Theorem 2.3], Haraux-Jendoubi [**61**, Theorem 3.1], and Chill-Fašangová [**19**]. Our formulation of the conditions is in the same spirit as Chill-Jendoubi [**21**, Theorem 2.3] (cf. also Chill-Fašangová [**19**]), which originates from Haraux-Jendoubi [**61**, Theorem 3.1]. One of our extra conditions is that there holds a stronger gradient inequality (see (3.38a) below). Our proof tools are Lemmas 2.2 and 2.3 of this chapter.

THEOREM 4.5. **(Convergence in gradient-like systems of wave type)** *Under the above situations (where conditions (i)–(iv) are satisfied), we let $u : [0,\infty) \to X$ be a solution of (3.32a) with the* **negative sign**. *Assume that*

$$\{\dot{u}(t) : t \geq 0\} \subset X, \quad \sup\{\|\dot{u}(t)\|_H : t \geq 0\} < \infty,$$

and that the orbit $O(u) = \{u(t) : t \geq 0\}$ is precompact in X.

Moreover, assume that there exists a Hilbert space $V \subset H$ with the following properties:

(i) There exist a cluster point $\varphi \in X$ of the orbit $O(u)$, a constant $\sigma > 0$ and a function $\phi \in \mathcal{G}_k$ such that

$$\phi(\mathcal{E}(w)) \leq \|\mathcal{E}'(w)\|_{V'} \quad \forall w \in X, \|w - \varphi\|_X < \sigma. \tag{3.38a}$$

(ii) There exists a constant $c_1 > 0$ such that

$$\|B(t, u(t), \dot{u}(t))\|_{V'}^2 \leq c_1 \cdot \langle B(t, u(t), \dot{u}(t)), \dot{u}(t)\rangle_{H\times H} \quad \forall t \geq 0. \tag{3.38b}$$

(iii) If $K : V' \to V$ denotes the duality map, then each of the linear operators $K \circ \mathcal{E}''(u(t)) : X \to V$ admits an extension to a bounded linear operator on H such that

$$c_1' := \sup\{\|K \circ \mathcal{E}''(u(t))\|_{L(H)} : t \geq 0\} < \infty. \tag{3.38c}$$

Then there holds

$$\lim_{t\to\infty} (\|\dot{u}(t)\|_H + \|u(t) - \varphi\|_X) = 0, \quad \mathcal{E}'(\varphi) = 0.$$

PROOF. The proof is very similar to the proof for Theorem 4.4, and it is divided into three steps.

Step 1. For a positive constant, whose smallness will be determined later, we consider the function

$$h(t) := -\frac{\kappa}{2}\|\dot{u}(t)\|_H^2 - \kappa\mathcal{E}(u(t)) - \alpha\langle \mathcal{E}'(u(t)), \dot{u}(t)\rangle_{V'\times V'} \quad (t \geq 0). \tag{3.39}$$

(Note that the difference between the function given by (3.39) and (3.34) is the form of the last two terms: In (3.34) we choose the positive sign and the scalar product is taken over the Hilbert space H, while in (3.39) we choose the negative sign and the scalar product is taken over the Hilbert space V'.)

By a similar manipulation as that in the proof of Theorem 4.4, we have, with $u \equiv u(t) \in X, v \equiv \dot{u}(t) \in X$, and $B \equiv B(t,u,v)$, that

$$h'(t) = \langle v, B\rangle + \alpha\|\mathcal{E}'(u)\|_{V'}^2 + \tilde{h}(t),$$

where

$$\tilde{h}(t) := -\kappa\langle v, g(t,u,v)\rangle + \alpha\langle \mathcal{E}'(u), \kappa B - g\rangle_{V'\times V'} - \alpha\langle \mathcal{E}''(u)v, v\rangle_{V'\times V'}.$$

Here and below, we use $\langle\cdot,\cdot\rangle$ and $\|\cdot\|$ to denote the scalar product and the norm on the Hilbert space H, respectively.

Using the Schwarz inequality combined with the elementary inequality $2|ab| \leq a^2 + b^2$ $(a,b \in \mathbb{R})$, we have that

$$\begin{aligned}&|\langle \kappa v, g(t,u,v)\rangle| \leq \|v\|\cdot g(t) \leq c\|v\|^2/2 + g(t)^2/c^2 \\ &\leq \langle B, v\rangle/2 + g(t)^2/c^2 \quad \text{(by (3.32b))}.\end{aligned}$$

Similarly,

$$|\langle \mathcal{E}'(u), \kappa B - g(t,u,v)\rangle_{V'\times V'}| \leq \|\mathcal{E}'(u)\|_{V'}\cdot(\|B\|_{V'} + \|g(t,u,v)\|_{V'}).$$

Thus, if $\delta > 0$ is the norm of the continuous injection $H \hookrightarrow V'$, then the estimate goes further as follows:

$$\begin{aligned}&\leq \|\mathcal{E}'(u)\|_{V'}\cdot(\|B\|_{V'} + \delta\|g(t,u,v)\|) \\ &\leq \|\mathcal{E}'(u)\|_{V'}\cdot(\|B\|_{V'} + \delta g(t)) \quad \text{(by (3.32c))} \\ &\leq (1/2)\|\mathcal{E}'(u)\|_{V'}^2 + \|B\|_{V'}^2 + \delta^2 g(t)^2 \\ &\leq (1/2)\|\mathcal{E}'(u)\|_{V'}^2 + c_1\langle B, v\rangle + \delta^2 g(t)^2 \quad \text{(by (3.38b))}.\end{aligned}$$

On the other hand, we have that

$$\begin{aligned}&|\langle \mathcal{E}''(u)v, v\rangle_{V'\times V'}| \leq \|K\circ\mathcal{E}''(u)\|_{L(H)}\cdot\|v\|^2 \\ &\leq c_1'\|v\|^2 \quad \text{(by (3.38c))} \\ &\leq (c_1'/c)\langle B, v\rangle \quad \text{(by (3.32b))}.\end{aligned}$$

Therefore,

$$\begin{aligned}|\tilde{h}(t)| \;\leq\; &(\alpha/2)\|\mathcal{E}'(u)\|_{V'}^2 + (1/2 + \alpha c_1 + \alpha c_1'/c)\langle B, v\rangle \\ &+(\alpha\delta^2 + 1/c^2)g(t)^2\end{aligned}$$

and thus,

$$\begin{aligned}h'(t) \;\geq\; &(1/2 - \alpha c_1 - \alpha c_1'/c)\cdot\langle B, v\rangle + (\alpha/2)\|\mathcal{E}'(u)\|_{V'}^2 \\ &-(\alpha\delta^2 + 1/c^2)g(t)^2.\end{aligned}$$

Consequently, if

$$0 < \alpha < c/(2c_1c + 2c_1'), \tag{3.40a}$$

then, since $\langle B, v\rangle \geq c\|v\|^2 \geq 0$ by (3.32b), we have (by recovering $u = u(t), v = \dot{u}(t)$) that

$$h'(t) \geq c_2\cdot(\|\dot{u}(t)\| + \|\mathcal{E}'(u(t))\|_{V'})^2 - c_3 g(t)^2 \quad \forall t \geq 0 \tag{3.40b}$$

with $c_2 := \frac{1}{4}\min\{\alpha, c - 2\alpha(cc_1 + c_1')\} > 0$ and $c_3 := \alpha\delta^2 + 1/c^2$. This implies that the new function

$$H(t) := h(t) + g_1(t) \quad (t \geq 0) \tag{3.40c}$$

with $g_1(t) := c_3 \int_t^\infty g(s)^2\, ds$ is a bounded, nondecreasing and absolutely continuous function, such that

$$H'(t) \geq c_2 \cdot (\|\dot{u}(t)\| + \|\mathcal{E}'(u(t))\|_{V'})^2 \quad \forall t \geq 0. \tag{3.40d}$$

Step 2. In this step we use (3.40d) to prove the convergence

$$\lim_{t\to\infty} \|\dot{u}(t)\| = 0.$$

Since the procedure is completely similar to that in the proof for Theorem 4.4, we omit its details.

Step 3. In this step we show the convergence $u(t) \to \varphi$ as $t \to \infty$ in X, where $\varphi \in X$ is the cluster point given in the theorem. For this purpose, we use the gradient inequality (3.38a) to find that there holds

$$\phi_1(\kappa\mathcal{E}(u)) \leq \|\mathcal{E}'(u)\|_{V'} \quad \forall u \in X, \|u - \varphi\|_X < \sigma, \tag{3.41a}$$

and

$$\phi_1(x + y) \leq k\phi_1(x) + \sqrt{|y|} \quad (x, y \in \mathbb{R}) \tag{3.41b}$$

with the new function $\phi_1 \in \mathcal{G}_k$ given by $\phi_1(s) := \phi(\kappa s)$ ($\forall s \in \mathbb{R}$). (Remember that $\kappa^2 = 1$.)

Consider the function H given by (3.40c) (see also (3.39) for the definition of h). As in the proof for Theorem 4.4, we have (mainly under the use of (3.41b)) that

$$\phi_1(H(t)) \leq k\phi_1(\kappa\mathcal{E}(u(t))) + \alpha\|\mathcal{E}'(u(t))\|_{V'} + (1 + \delta)\|\dot{u}(t)\| + \sqrt{g_1(t)}.$$

Let

$$W := \{t \geq 0 : \|u(t) - \varphi\|_X < \sigma\}.$$

Then, for $t \in W$ we have by the gradient inequality (3.41a) that $\phi_1(\kappa\mathcal{E}(u(t))) \leq \|\mathcal{E}'(u(t))\|_{V'}$. Hence, we have for all $t \in W$ that

$$\phi_1(H(t)) \leq (k + \alpha)\|\mathcal{E}'(u(t))\|_{V'} + (1 + \delta)\|\dot{u}(t)\| + \sqrt{g_1(t)},$$

and thus, by using (3.40d) and noting $k \geq 1$, we find that

$$\phi_1(H(t)) \leq ((k + \alpha + \delta)/\sqrt{c_2})\sqrt{H'(t)} + \sqrt{g_1(t)} \quad \forall t \in \Lambda.$$

Consequently, with the function $\phi_2 := \sqrt{c_2}/(k + \alpha + \delta)\phi_1 \in \mathcal{G}$, we have that

$$\phi_2(H(t)) \leq \sqrt{H'(t)} + \sqrt{g_1(t)} \quad \forall t \in W.$$

By (3.32c), $\sqrt{g_1} \in L^1(\mathbb{R}_+)$. Therefore, the function H satisfies the inequalities (3.8) of Lemma 2.2, and thus $\sqrt{H'(\cdot)} \in L^1(W)$ by Lemma 2.2 of this chapter. This implies by (3.40d) that

$$\int_W \|\dot{u}(t)\|\, dt < \infty.$$

A similar argument as that in the proof for Theorem 4.4 (under the use of Lemma 2.3) finally yields the desired result $\|u(t) - \varphi\|_X \to 0$ as $t \to \infty$ and $\mathcal{E}'(\varphi) = 0$. This completes the proof. □

As consequences, we have the following convergence results for analytic maps.

COROLLARY 4.6. *(i) Both Theorems 4.2 (ii) and Theorem 4.4 apply to the case where the gradient map $\mathcal{E}' : X \to H$ is analytic in the sense that hypotheses* **(H1)–(H3)** *in §4 of Chapter 2 are satisfied.*

(ii) Theorem 4.5 applies to the case where X is itself a Hilbert space and $\mathcal{E} : X \to \mathbb{R}$ is analytic such that, for each $w \in X$, the operator $\mathcal{E}''(w) : X \to X'$ is a linear Fredholm operator of index zero.

PROOF. For case (i) we use Theorem 4.2 of Chapter 2 to conclude that the gradient map $\mathcal{E}' : X \to H$ locally satisfies a gradient inequality of LS-type, while for case (ii) we take $V := X$ and use Theorem 4.5 of Chapter 2 to conclude that the gradient map $\mathcal{E}' : X \to V' = X'$ locally satisfies a gradient inequality of LS-type. □

REMARK 4.7. We look back at the conditions posed in both Theorems 4.4 and 4.5.

(a) It is easily verified that condition (3.33a) for Theorem 4.4 will be satisfied by the semilinear gradient maps $\mathcal{E}'$ of the form $\mathcal{E}'(u) = Au + f(\cdot, u)$, where A is a positive definite operator. Such semilinear gradient maps are met in many practical application situations.

(b) There is a big difference between the conditions on the damping term B for both theorems. The boundedness condition (3.33b) for Theorem 4.4 given in the H norm yields in general the constraint that the damping term $B(t, u, \dot{u})$ is bounded. However, the boundedness condition (3.38b) for Theorem 4.5 is given in the V' norm, which is in general much weaker than the H norm. Thus, Theorem 4.5 is applicable to the case where the damping term B is *unbounded*.

(c) The biggest difference between the conditions for Theorems 4.4 and 4.5 is the validity of gradient inequality. For Theorem 4.3 we need a very mild gradient inequality (3.33c) using the H norm which, as shown in Chapter 2, has been satisfied by many gradient maps. However, the gradient inequality (3.38a) using the V' norm is certainly very sharp. For instance, if we take $\mathcal{E}'$ to be the semilinear elliptic operator $\mathcal{E}'(u) := -\Delta u + f(\cdot, u)$ so that the resulting equation is the semilinear wave equation, then the analyticity requirement of the corresponding functional $\mathcal{E} : X \to \mathbb{R}$ yields in general a very strong constraint on the growth of the nonlinearity $f(\cdot, u)$ (e.g., $f(\cdot, u)$ being a polynomial in u). Hence, the result in Theorem 4.5 is not optimal for the usual semilinear wave equations. Fortunately, in the next chapter we can prove a gradient inequality (Theorem 1.1 of Chapter 4) that will be applicable to such a semilinear case and thus the results for the corresponding gradient-like systems of wave type can be improved in a very optimal way; see Theorem 1.2 of Chapter 4.

4C: Gradient inequalities and convergence rate. For simplicity, we consider the gradient system

$$\dot{u} = -\mathcal{E}'(u) \tag{3.42}$$

over the Hilbert space H. We assume that U is a subset of X for which there exists some constant $\gamma > 0$ such that

$$0 \le \mathcal{E}(v) \le \gamma \quad \forall v \in U. \tag{3.43a}$$

Moreover, we assume that $\mathcal{E}'$ satisfies a gradient inequality in U

$$\phi(\mathcal{E}(v)) \le \|\mathcal{E}'(v)\|_H \quad \forall v \in U \tag{3.43b}$$

with a function $\phi \in \mathcal{G}$ such that ϕ is strictly positive on $(0, \gamma]$. It follows that the integrals

$$(0, \gamma] \ni x \mapsto \int_x^\gamma \frac{ds}{\phi(s)^2}$$

well define an absolutely continuous and strictly increasing function on $(0, \gamma]$. Therefore, there exists a unique function $g : \mathbb{R}_+ \to (0, \gamma]$ such that

$$t = \int_{g(t)}^\gamma \frac{ds}{\phi(s)^2} \quad \forall t \ge 0. \tag{3.44a}$$

We have that $g(0) = \gamma$ and $g(t) \to 0$ as $t \to \infty$. Let

$$\Phi(x) := \int_0^x \frac{ds}{\phi(s)} \quad (x \in \mathbb{R}). \tag{3.44b}$$

This is an absolutely continuous and strictly increasing function.

THEOREM 4.8. *Under the above situations we let $u : \mathbb{R} \to X$ be a solution of the gradient system (3.42), and we assume that the orbit of u is contained in U, i.e., $O(u) \subset U$. Then there exists some $u_\infty \in H$ such that*

$$\|u(t) - u_\infty\|_H \le \Phi(g(t)) \quad \forall t \ge 0. \tag{3.45}$$

Roughly speaking, Theorem 4.8 says that the convergence rate of a solution of the gradient system (3.42) can be estimated as above by merely using the function ϕ, which is involved in the corresponding gradient inequality in its orbit. Moreover, the convergence rate can be uniformly estimated for these solutions that share a common function ϕ.

Proof of Theorem 4.8. For the function $H(t) := \mathcal{E}(u(t))$ $(t \ge 0)$ we have that

$$-\dot{H}(t) = \|\dot{u}(t)\|^2 = \|\mathcal{E}'(u(t))\|^2$$

for all $t \ge 0$. It follows from the assumption $O(u) \subset U$ combined with the boundedness condition (3.43a) that the limit $a := \lim_{t\to\infty} H(t)$ exists and $a \ge 0$. Moreover, for a.e. $t \ge 0$, we have that

$$-\frac{d}{dt}\Phi(H(t)) = \|\dot{u}(t)\| \cdot \frac{\|\mathcal{E}'(u(t))\|}{\phi(\mathcal{E}(u(t)))}.$$

Since each $u(t) \in U$, we are allowed to apply the gradient inequality (3.43b) to the elements $v = u(t)$. It yields that

$$-\dot{H}(t) \geq \phi(H(t))^2 \quad \forall t \geq 0, \tag{3.46a}$$

and

$$-\frac{d}{dt}\Phi(H(t)) \geq \|\dot{u}(t)\| \quad \text{a.e.} \quad t \in \mathbb{R}_+. \tag{3.46b}$$

Taking integration in (3.46b) and noting that $\Phi(a) \geq 0$, since $a \geq 0$, we find that

$$\int_t^\infty \|\dot{u}(s)\| \, ds \leq \Phi(H(t)) \quad \forall t \geq 0. \tag{3.46c}$$

As seen before, (3.46c) has already implied the convergence $u(t) \to u_\infty \in H$ in H as $t \to \infty$ as well as the estimate

$$\|u(t) - u_\infty\| \leq \Phi(H(t)) \quad \forall t \geq 0. \tag{3.46d}$$

Clearly, if there is some $t_0 \geq 0$ such that $H(t_0) = 0$, then we must have $u(t) \equiv u(0)$ and $u(0)$ is a critical point of $\mathcal{E}$. In this case, we always have $u(t) - u_\infty \equiv 0$ and thus (3.45) follows. Because of this, we generally assume that $H(t) > 0$ for all $t \geq 0$. Then we have that $H(t) \in (0, \gamma]$ and thus $\phi(H(t)) > 0$ for all $t \geq 0$, since the function $\phi \in \mathcal{G}$ is strictly positive on $(0, \gamma]$. Hence, we can use (3.46a) to find that

$$-\frac{\dot{H}(t)}{\phi(H(t))^2} \geq 1 \quad (t \geq 0).$$

An integration of the above inequality over any finite interval $(0, t)$ yields that

$$t \leq -\int_0^t \frac{\dot{H}(s)}{\phi(H(s))^2} \, ds = \int_{H(t)}^{H(0)} \frac{ds}{\phi(s)^2} \leq \int_{H(t)}^{\gamma} \frac{ds}{\phi(s)^2},$$

since $H(0) \leq \gamma$. By the definition of the function g, we have that

$$\int_{g(t)}^{\gamma} \frac{ds}{\phi(s)^2} = t \leq \int_{H(t)}^{\gamma} \frac{ds}{\phi(s)^2}$$

and thus $H(t) \leq g(t)$ for all $t \geq 0$. Using this in (3.46d), we obtain the desired result in (3.45). □

Below we compute two concrete examples.

EXAMPLE 4.9. (i) Let $\theta \in [1/2, 1)$ and $c > 0$ be a constant. For the increasing function $\phi(x) = c \cdot x^\theta$ $(x \geq 0)$, which corresponds to a gradient inequality of LS-type, we have for all $x > 0$ that

$$\Phi(x) = \int_0^x \frac{ds}{\phi(s)} = \frac{1}{c(1-\theta)} \cdot x^{1-\theta}.$$

For any $\gamma > 0$, the function $g : \mathbb{R}_+ \to [0, \gamma)$ given by (3.44a) is determined by the equation

$$c^2 t = c^2 \int_{g(t)}^{\gamma} \frac{ds}{\phi(s)^2} = \int_{g(t)}^{\gamma} x^{-2\theta}\, dx \quad (t \geq 0).$$

If $\theta = 1/2$, then we have that $g(t) = \gamma \cdot \exp(-c^2 t)$ for all $t \geq 0$ and thus

$$\Phi(g(t)) = (2\sqrt{\gamma}/c) \exp(-c^2 t/2) \quad \forall t \geq 0.$$

If $\theta > 1/2$, then an elementary calculation yields that

$$\Phi(g(t)) = \frac{1}{c(1-\theta)} \cdot [(2\theta - 1)c^2 t + \gamma^{1-2\theta}]^{-\left(\frac{1-\theta}{2\theta-1}\right)}$$

for all $t \geq 0$. Thus, the convergence rate is exponentially rapid if the gradient inequality (3.44b) is of LS-type with the best exponent $\theta = 1/2$, and the convergence rate is in general only polynomially or subpolynomially rapid, depending on the exponent θ. The smaller the exponent θ is, the larger the convergence rate is. This assertion has been obtained by Haraux and Jendoubi in [**60**].

(ii) For the next calculation we fix two numbers $c > 0$, $\alpha > 1$ and we let $\gamma < 1$ be a given positive constant. We take ϕ to be a function in $\mathcal{G}$ whose part on $[0, \gamma]$ is given by $\phi(0) := 0$ and

$$\phi(x) := c \cdot x(-\ln x)^{\alpha} \quad \forall x \in (0, \gamma].$$

Correspondingly, we have that $\Phi(x) = (-\ln x)^{1-\alpha}/(c(\alpha-1))$ for all $x \in (0, \gamma]$ and the function $g \leq \gamma$ is determined by the equation

$$(*) \qquad c^2 t = \int_{g(t)}^{\gamma} s^{-2}(-\ln s)^{-2\alpha}\, ds \quad (t \geq 0).$$

Using the estimate

$$(-\ln s)^{-2\alpha} \leq (-\ln \gamma)^{-2\alpha}$$

for all $s \in (0, \gamma]$, we find that

$$(c^2(-\ln \gamma)^{2\alpha}) \cdot t \leq \int_{g(t)}^{\gamma} s^{-2}\, ds = \frac{1}{g(t)} - \frac{1}{\gamma}$$

and thus

$$g(t) \leq 1/[1 + (c^2 \gamma(-\ln \gamma)^{2\alpha}) \cdot t] \quad \forall t \geq 0$$

by noting that $\gamma < 1$. It follows that

$$(**) \qquad \Phi(g(t)) \leq \frac{1}{c(\alpha - 1)} \cdot \frac{1}{[\ln(1 + (c^2\gamma(-\ln \gamma)^{2\alpha}) \cdot t)]^{\alpha-1}} \quad \forall t \geq 0.$$

Therefore, the convergence rate is logarithmic and thus very slow.

We compare this assertion with the following concrete one-dimensional example. A special class of functions on $\mathbb{R}$ that satisfy gradient inequalities of the above type is the C^{∞} functions given by

$$f_{\alpha}(x) := \exp(-|x|^{-\frac{1}{\alpha-1}}) \quad (x \in \mathbb{R}),$$

where the constants $\alpha > 1$. We have, with $\phi_\alpha(x) := x(-\ln x)^\alpha/(\alpha-1)$, that

$$\phi_\alpha(f_\alpha(x)) = f'_\alpha(x) \quad (x \in \mathbb{R}).$$

Now the above result asserts that the convergence rate of solutions $x(t)$ of the one-dimensional equation $\dot{x} = -f'_\alpha(x)$ with $x(0) > 0$ to the unique critical point $x = 0$ is given by

$$0 \le x(t) \le \Phi(g(t)) \quad \forall t \ge 0,$$

where $\Phi(y) := (-\ln y)^\alpha$ for all $y > 0$ and $g(t)$ is determined as in $(*)$, under the choice of the constant $\gamma = f_\alpha(x(0))$. On the other hand, a careful but elementary manipulation shows that

$$x(t) = \Phi(g(t)) \quad \forall t \ge 0.$$

Therefore, the convergence rate given by the above general theory for this case is exact.

4D: Gradient inequalities and Lyapunov stability of ground states. As in **4C**, we consider the gradient system

$$\dot{u} = -\mathcal{E}'(u), \quad u(0) = u_0 \in H \tag{3.47a}$$

over the Hilbert space H. As in the beginning of this section, $\mathcal{E}' : X \subset H \to H$ is a continuous gradient map.

Let $U \subset H$ (not necessarily contained in X) be a given set such that the intersection $U \cap X$ is closed in X, i.e.,

$$U \cap X \quad \text{is closed in} \quad X. \tag{3.47b}$$

We assume that $\varphi \in U \cap X$ is a *local ground state* for the C^1 functional $\mathcal{E} : U \cap X \to \mathbb{R}$; i.e., φ is a critical point of $\mathcal{E}$ and $\mathcal{E} : U \cap X \to \mathbb{R}$ attains its local minimum at φ :

$$\mathcal{E}'(\varphi) = 0, \quad \mathcal{E}(\varphi) \le \mathcal{E}(u) \quad \forall u \in U \cap X. \tag{3.47c}$$

We are interested in the stability of the local ground state φ as a stationary solution of the corresponding gradient system (3.47a). For this purpose, we assume that the solutions of the gradient system (3.47a) obey the following properties (i)–(iii):

(i) **(Existence and continuous dependence)** For each initial value $u_0 \in U$, the gradient system (3.47a) has a unique and global solution $u(\cdot; u_0)$ such that

$$u(\cdot; u_0) \in C([0,\infty); H) \cap C^1((0,\infty); H) \cap C((0,\infty); U \cap X). \tag{3.48a}$$

Moreover, the solutions are dependent on the initial data in the sense that there exists a constant $c_1 > 0$ such that

$$\|u(1; u_0) - u(1; v_0)\|_H \le c_1 \cdot \|u_0 - v_0\|_H \quad \forall u_0, v_0 \in U. \tag{3.48b}$$

(ii) **(Uniform compactness of solutions)** There exists a compact subset K of X such that

$$\{u(t; u_0) : t \ge 1\} \subset K \quad \forall u_0 \in U. \tag{3.48c}$$

(iii) **(Validity of gradient inequality)** The gradient map $\mathcal{E}' : X \to H$ satisfies a gradient inequality in K; i.e., there exists some function $\phi \in \mathcal{G}$ such that

$$\phi(\mathcal{E}(v)) \leq \|\mathcal{E}'(v)\|_H \quad \forall v \in K. \tag{3.48d}$$

Note that in (i) we have required that the points $u(t; u_0)$ for all $t > 0$ are in the intersection $U \cap X$. This implies that the solutions $u(t; u_0)$ with $u_0 \in U$ obey a smoothing effect for all $t > 0$. We will see in the next chapter that the above properties in (i)–(ii) for the solutions will be satisfied in many semilinear gradient systems. However, the property in (iii) for the validity of gradient inequality is just as crucial, since one sometimes needs the analyticity of the gradient maps, as shown in the previous chapter.

Below we will show that the validity of a gradient inequality of the above type implies the following stability of the ground state φ : If the initial value $u_0 \in U$ is sufficiently close to φ in the H norm (not necessarily in the X norm!), then the solution $u(\cdot; u_0)$ of the gradient system (3.47a) remains near φ in the X norm (!) for all times $t \geq 1$. More precisely, our result reads as follows.

THEOREM 4.10. **(Stability of ground state)**
Assume that the above conditions (i)–(iii) are satisfied, and let $\varphi \in U \cap X$ be a local ground state satisfying (3.47c). Then φ is Lyapunov stable in the following sense: For any given $\varepsilon > 0$, there exists a constant $\delta \equiv \delta(\varepsilon) > 0$ such that

$$u_0 \in U, \|u_0 - \varphi\|_H < \delta \quad \Longrightarrow \quad \sup_{t \geq 1} \|u(t; u_0) - \varphi\|_X < \varepsilon, \tag{3.49a}$$

where $u(\cdot; u_0)$ is the solution of the gradient system (3.47a) with the initial value $u(0; u_0) = u_0$.

Moreover, if φ is an isolated critical point of $\mathcal{E}$ in $U \cap X$, then φ is uniformly asymptotically stable in the sense that there exists a constant $\delta_0 > 0$ such that

$$\|u(t; u_0) - \varphi\|_X \to 0 \quad \text{as} \quad t \to \infty \tag{3.49b}$$

uniformly for all initial data $u_0 \in U$ with $\|u_0 - \varphi\|_H < \delta_0$.

REMARK 4.11. Note that the validity of the gradient inequality in K ensures particularly that there holds the same gradient inequality in the section $\{u(t; u_0) : t \geq 1\}$, where the initial value $u_0 \in U$ is arbitrary, since $\{u(t; u_0) : t \geq 1\}$ is contained in the compact set $K \subset X$ by assumption (ii). As seen before (see, e.g., Theorem 4.2 of this chapter), this implies that there exists some critical point $\varphi(u_0) \in X$ of $\mathcal{E}$ such that $\|u(t; u_0) - \varphi(u_0)\|_X \to 0$ as $t \to \infty$.

One implication of the Lyapunov stability given in Theorem 4.10 for these limits $\{\varphi(u_0) : u_0 \in U\}$ is that there holds $\|\varphi(u_0) - \varphi\|_X \to 0$ whenever $\|u_0 - \varphi\|_H \to 0$.

Proof of Theorem 4.10. Part 1: The Lyapunov stability. Let

$$\Phi(x) := \int_{\mathcal{E}(\varphi)}^{x} \frac{ds}{\phi(s)} \quad (x \in \mathbb{R}),$$

where $\phi \in \mathcal{G}$ is the function such that the gradient inequality (3.48d) holds. Φ is a nondecreasing and absolutely continuous function.

Consider a solution u of (3.47a) with an initial value $u(0) = u_0 \in U$. By (i) we have that $u(t) \in U \cap X$ for all $t > 0$. This implies that

$$\mathcal{E}(u(t)) \geq \mathcal{E}(\varphi) \quad (t > 0), \tag{3.50a}$$

since φ is a local ground state. Moreover, since the section $\{u(t) : t \geq 1\}$ is contained in K by assumption (ii), we have by (3.48d) that

$$\phi(\mathcal{E}(u(t))) \leq \|\mathcal{E}'(u(t))\|_H \quad \forall t \geq 1. \tag{3.50b}$$

Set

$$H(t) := \mathcal{E}(u(t)) \quad (t \geq 0).$$

Then, as seen in the proof of Theorem 4.8 (cf. (3.46b)), the above gradient inequality (3.50b) implies that

$$-\frac{d}{dt}\Phi(H(t)) \geq \|\dot{u}(t)\|_H \quad a.e. \quad t \in [1, \infty). \tag{3.50c}$$

Therefore, by integrating (3.50c), we obtain that

$$\int_1^t \|\dot{u}(s)\|_H \, ds \leq \Phi(H(1)) - \Phi(H(t))$$

for all $t \geq 1$. On the other hand, we have by (3.50a) that $\Phi(H(t)) \geq \Phi(\mathcal{E}(\varphi)) = 0$ for all $t \geq 1$. Hence, by using the inequality $\|u(t) - u(1)\|_H \leq \int_1^t \|\dot{u}(s)\|_H \, ds$ $(t \geq 1)$, we find from above that

$$\sup_{t \geq 1} \|u(t) - u(1)\|_H \leq \Phi(\mathcal{E}(u(1))). \tag{3.51a}$$

To continue, we use (3.48b) to find that

$$\|u(1) - \varphi\|_H \leq c_1 \cdot \|u_0 - \varphi\|_H, \tag{3.51b}$$

since φ is a stationary solution of (3.47a). Combining this with (3.51a), we obtain that

$$\sup_{t \geq 1} \|u(t) - \varphi\|_H \leq \Phi(\mathcal{E}(u(1))) + c_1\|u_0 - \varphi\|_H. \tag{3.51c}$$

Now we are in the position to prove the desired uniform stability of the local ground state φ. Assume, conversely, that φ is not Lyapunov stable in the sense given in the theorem; then there exists a constant $\varepsilon_0 > 0$ and a sequence of solutions $\{u_n : n \in \mathbb{N}\}$ of the gradient system (3.47a) with initial data $u_n(0) \in U$ such that

$$\|u_n(0) - \varphi\|_H < 1/n, \tag{3.52a}$$

but there exists a sequence $\{t_n : n \in \mathbb{N}\} \subset [1, \infty)$ such that

$$\|u_n(t_n) - \varphi\|_X \geq \varepsilon_0 > 0. \tag{3.52b}$$

Applying (3.51b,c) to the solutions u_n and using (3.52a), we find for all $n \in \mathbb{N}$ that

$$\|u_n(1) - \varphi\|_H \leq c_1 \|u_n(0) - \varphi\|_H < c_1/n, \tag{3.52c}$$

and

$$\sup_{t \geq 1} \|u_n(t) - \varphi\|_H \leq \Phi(\mathcal{E}(u_n(1))) + c_1/n. \tag{3.52d}$$

On the one hand, we have that $\|u_n(1) - \varphi\|_H \to 0$ as $n \to \infty$. On the other hand, by assumption (ii), the sequence $\{u_n(1) : n \in \mathbb{N}\}$ is contained in the compact subset K of X and thus it is precompact in X. Therefore, the convergence $u_n(1) \to \varphi$ as $n \to \infty$ in the H norm implies that φ is the unique cluster point of the precompact sequence $\{u_n(1) : n \in \mathbb{N}\}$ in X. Hence, we must have that

$$\lim_{n \to \infty} \|u_n(1) - \varphi\|_X = 0.$$

This implies that

$$\lim_{n \to \infty} \Phi(\mathcal{E}(u_n(1))) = \Phi(\mathcal{E}(\varphi)) = 0,$$

since both $\mathcal{E}$ and Φ are continuous. As a consequence, we have by (3.52d) that

$$\lim_{n \to \infty} \sup_{t \geq 1} \|u_n(t) - \varphi\|_H = 0.$$

In particular,

$$\lim_{n \to \infty} \|u_n(t_n) - \varphi\|_H = 0. \tag{3.52e}$$

However, this implies a contradiction. In fact, since the sequence $\{u_n(t_n)\}$ is contained in the compact subset $K \subset X$ by assumption (ii) (see (3.48c)), the convergence in (3.52e) implies that φ is the unique cluster point of the precompact sequence $\{u_n(t_n)\}$ in $X \subset H$ and thus we must have that $\|u_n(t_n) - \varphi\|_X \to 0$ as $n \to \infty$, contradicting (3.52b). This finishes the proof of the Lyapunov stability of φ. □

Proof of Theorem 4.10. Part 2: The uniform asymptotic stability. We assume that φ is an isolated critical point in $U \cap X$. It follows that there exists a number $\varepsilon_1 > 0$ such that the subset

$$U_1 := \{w \in U \cap X : \|w - \varphi\|_X \leq \varepsilon_1\} \tag{3.53a}$$

contains only one critical point of $\mathcal{E}$, namely, the isolated equilibrium φ. Note that U_1 is closed in X, since the intersection $U \cap X$ has been assumed to be closed in X; see (3.47b).

Next we use the Lyapunov stability of φ, which we have just established. It yields that there exists some constant $\delta_1 > 0$ such that

$$\sup_{t \geq 1} \|u(t; u_0) - \varphi\|_X < \varepsilon_1 \tag{3.53b}$$

for all initial data $u_0 \in U_2$, where U_2 is given by

$$U_2 := \{w \in U : \|w - \varphi\|_H < \delta_1\}. \tag{3.53c}$$

We will show that there holds the uniform convergence

$$\lim_{t\to\infty} \sup_{u_0 \in U_2} \|u(t;u_0) - \varphi\|_X = 0. \tag{3.53d}$$

Our tool for proving this is Theorem 4.8. In order to apply Theorem 4.8, we must first construct a new function $\tilde{\phi} \in \mathcal{G}$ such that the gradient inequality (3.48d) holds true in some subset K_1 of K and obeys the new property that $\tilde{\phi}(x) > 0$ for all $x \neq \mathcal{E}(\varphi)$.

Now we pursue the construction of the new function $\tilde{\phi} \in \mathcal{G}$. To this end, we consider the set

$$K_1 := \{u(t;u_0) : u_0 \in U_2, t \geq 1\} \subset K. \tag{3.54a}$$

We have $K_1 \subset U_1$ by (3.53b). Moreover, since $K_1 \subset K$ and K is compact in X, the closure $\bar{K}_1$ of K_1 in X is compact. Therefore, by the closedness of the set U_1 in X, we have that

$$\bar{K}_1 \subset U_1. \tag{3.54b}$$

As seen in Remark 4.11, the gradient inequality (3.48d) implies the convergence $\|u(t;u_0) - \varphi(u_0)\|_X \to 0$ as $t \to \infty$ with some critical point $\varphi(u_0)$. By (3.54b) we have that $\varphi(u_0) \in U_1$ for all $u_0 \in U_2$. Since φ is the unique critical point in U_1, we must have that $\varphi(u_0) = \varphi$ and thus

$$\lim_{t\to\infty} \|u(t;u_0) - \varphi\|_X = 0 \quad \forall u_0 \in U_2. \tag{3.54c}$$

This implies that all of the number sets

$$I(u_0) := \{\mathcal{E}(u(t;u_0)) : t \geq 1\} \cup \{\mathcal{E}(\varphi)\} \quad (u_0 \in U_2)$$

are path-connected and have the point $\mathcal{E}(\varphi)$ in common. It follows that

$$\mathcal{E}(\bar{K}_1) = (\cup_{u_0 \in U_2} I(u_0))^-$$

is path-connected and closed and thus there exists some $r \geq \mathcal{E}(\varphi)$ such that

$$\mathcal{E}(\bar{K}_1) = [\mathcal{E}(\varphi), r]. \tag{3.54d}$$

Without loss of generality, we may assume that $r > \mathcal{E}(\varphi)$.

We define a function $\tilde{\phi} : \mathbb{R} \to \mathbb{R}$ by setting $\tilde{\phi}(x) := 1$ for all $x \notin [\mathcal{E}(\varphi), r]$ and

$$\tilde{\phi}(x) := \inf\{\|\mathcal{E}'(w)\|_H : w \in \bar{K}_1, \mathcal{E}(w) = x\} \quad \forall x \in [\mathcal{E}(\varphi), r].$$

The new function $\tilde{\phi}$ is lower semicontinuous on $[\mathcal{E}(\varphi), r]$, and such that

$$\tilde{\phi}(\mathcal{E}(w)) \leq \|\mathcal{E}'(w)\|_H \quad \forall w \in K_1. \tag{3.55}$$

Moreover, we have that $\tilde{\phi}(x) \geq \phi(x)$ for all $x \in [\mathcal{E}(\varphi), r]$, where $\phi \in \mathcal{G}$ is the function satisfying the gradient inequality (3.48d). This implies that $\tilde{\phi} \in \mathcal{G}$. Furthermore, we have that $\tilde{\phi}(x) > 0$ for all $x \in (\mathcal{E}(\varphi), r]$. If there were some $x_0 \in (\mathcal{E}(\varphi), r]$ such that $\tilde{\phi}(x_0) = 0$, then the compactness of $\bar{K}_1$ implies that there exists some $w_0 \in \bar{K}_1 \subset U_1$ such that $\mathcal{E}'(w_0) = 0$ and

$\mathcal{E}(w_0) = x_0 > \mathcal{E}(\varphi)$. This implies that $w_0 \neq \varphi$, contradicting the fact that φ is the unique critical point of $\mathcal{E}$ in U_1.

With the new function $\tilde{\phi} \in \mathcal{G}$, we define the function

$$\tilde{\Phi}(x) := \int_{\mathcal{E}(\varphi)}^{x} \frac{ds}{\tilde{\phi}(s)} \quad (x \in \mathbb{R}).$$

Consider a solution $u(\cdot; u_0)$ of the gradient system (3.47a) with an initial value $u_0 \in U_2$. Then, by an appropriate variant of Theorem 4.8, we find that

$$\|u(t; u_0) - \varphi\|_H \leq \tilde{\Phi}(g(t)) \quad \forall t \geq 1 \tag{3.56a}$$

with the function $g : [1, \infty) \to (\mathcal{E}(\varphi), r]$ defined by the equation

$$t - 1 = \int_{g(t)}^{r} \frac{ds}{\tilde{\phi}(s)^2} \quad (t \geq 1);$$

cf. (3.44a). Clearly, we have that $g(t) \to \mathcal{E}(\varphi)$ as $t \to \infty$ and thus

$$\tilde{\Phi}(g(t)) \to 0 \quad \text{as} \quad t \to \infty. \tag{3.56b}$$

Assume, conversely, that (3.53d) is false; i.e., the convergence $\|u(t; u_0) - \varphi\|_X \to 0$ as $t \to \infty$ is not uniform for all $u_0 \in U_2$. Then there exists a constant $\varepsilon_2 > 0$ such that, for each $n \in \mathbb{N}$, there exists a solution $u_n(\cdot)$ of the gradient system (3.47a) with an initial value $u_n(0) \in U_2$ such that

$$\|u_n(n) - \varphi\|_X \geq \varepsilon_2 > 0. \tag{3.57}$$

By applying (3.56a) to each solution u_n, we find that

$$\|u_n(n) - \varphi\|_H \leq \tilde{\Phi}(g(n)) \to 0 \quad \text{as} \quad n \to \infty.$$

This implies that the sequence $\{u_n(n) : n \in \mathbb{N}\} \subset K$, which is precompact in X by the compactness of K in X, has only one cluster point φ in X and thus we must have that $\|u_n(n) - \varphi\|_X \to 0$ as $n \to \infty$, contradicting (3.57). This completes the proof for the uniform asymptotic stability. □

Applications of Theorem 4.10 to semilinear gradient systems will be given in the next chapter.

5. Convergence in variational problems

Let $\Omega \subset \mathbb{R}^N$ be a bounded domain with a Lipschitz boundary $\partial\Omega$, and let

$$H := L^2(\Omega; \mathbb{R}^m) \quad (m \in \mathbb{N}). \tag{3.58a}$$

We consider the functional

$$\mathcal{E}_F(u) := \int_{\Omega} F(x, u, Du)\, dx \quad (u \in C^1(\Omega; \mathbb{R}^m)), \tag{3.58b}$$

where the Carathéodory function $F : \bar{\Omega} \times \mathbb{R}^m \times \mathbb{R}^{mn} \to \mathbb{R}$ satisfies the properties (f1)–(f2) imposed in §5 of Chapter 2:

(f1) (**Uniform analyticity**) F is uniformly analytic in the sense that there exists a neighborhood $V \subset \mathbb{C}^m \times \mathbb{C}^{mn}$ of the real plane $\mathbb{R}^m \times \mathbb{R}^{mn}$

and an extension of F, denoted by $\tilde{F}$, such that $\tilde{F} : \bar{\Omega} \times V \to \mathbb{C}$ is locally bounded in $\bar{\Omega} \times V$, and $\tilde{F}(x, u, \xi)$ is analytic in both u and ξ variables.

(f2) (**Uniform ellipticity condition**) For each $R > 0$, there exists a constant $c \equiv c(R) > 0$ such that for all $(u, \xi) \in \mathbb{R}^m \times \mathbb{R}^{mn}$ with $|u| + |\xi| < R$, there holds

$$F''_{\xi\xi}(x, u, \xi)(v, \eta)^2 \geq c \cdot |v|^2 \cdot |\eta|^2 \quad \forall (x, v, \eta) \in \Omega \times \mathbb{R}^n \times \mathbb{R}^{nm}.$$

Here $F''_{\xi\xi}$ is the Hessian matrix of F with respect to the ξ variable.

Let p be such that $p \geq 2, \quad p > N$, and set

$$X_p := \{u \in W^{2,p}(\Omega; \mathbb{R}^m) \quad \text{plus some boundary condition}\}. \tag{3.58c}$$

Here the assigned boundary conditions are either the Dirichlet boundary condition, the Neumann boundary condition, or the periodic boundary condition. As ensured by Theorem 5.2 of Chapter 2, $\mathcal{E}'_F$ satisfies a gradient inequality of LS-type near each critical point of $\mathcal{E}_F$. More precisely, for each $\varphi \in X_p$, there are positive constants c, σ and $\theta \in [1/2, 1)$ such that

$$c \cdot |\mathcal{E}_F(u) - \mathcal{E}_F(\varphi)|^\theta \leq \|\mathcal{E}'_F(u)\|_{W^{-1,2}}$$

for all $u \in X_p$ with $\|u - \varphi\|_{X_p} < \sigma$.

Under the above settings, we consider the following first-order gradient-like system in the Hilbert space $H = L^2$:

$$\begin{cases} \dot{u} = -\mathcal{E}'_F(u) + g(t, u), \quad t \geq 0, \\ u \equiv u(t) \in L^2, \dot{u}(t) \in L^2, \end{cases} \tag{3.59a}$$

where the time-dependent perturbation $g : \mathbb{R}_+ \times X_p \to L^2$ is a continuous map and satisfies the following smallness condition:

$$\begin{cases} \|g(s, u)\|_{L^2} \leq g(s) \quad \forall s \geq 0, u \in X_p; \\ \displaystyle\int_0^\infty \rho(t) g(t)^2 \, dt + \int_0^\infty \frac{1}{\rho(t)} \, dt < \infty \\ \text{with some increasing positive function} \quad \rho. \end{cases} \tag{3.59b}$$

Moreover, the second-order gradient-like systems of *elliptic type* in $H = L^2$ have the form:

$$\begin{cases} \ddot{u} + \kappa B(t, u, \dot{u}) = \mathcal{E}'_F(u) + g(t, u, \dot{u}), \quad t \geq 0, \\ u \equiv u(t) \in L^2, \dot{u} \equiv \dot{u}(t) \in L^2, \ddot{u} \equiv \ddot{u}(t) \in L^2, \end{cases} \tag{3.60a}$$

for which the following conditions (i)–(iii) are satisfied:

(i) The constant κ is either equal to 1 or -1.

(ii) $B : \mathbb{R}_+ \times X_p \times L^2 \to L^2$ is a continuous map that is coercive and bounded in the sense that there exists a constant $c \in (0, 1)$ such that

$$\begin{cases} \langle B(t, u, v), v \rangle_{L^2 \times L^2} \geq c \cdot \|v\|^2_{L^2} \\ \|B(t, u, v)\|_{L^2} \leq (1/c) \cdot \|v\|_{L^2} \end{cases} \quad \forall (t, u, v) \in \mathbb{R}_+ \times X_p \times L^2. \tag{3.60b}$$

(iii) The perturbation $g : \mathbb{R}_+ \times X_p \times L^2 \to L^2$ is a continuous map and satisfies the following smallness condition:

$$\begin{cases} \sup_{u \in X_p, v \in L^2} \|g(t,u,v)\|_{L^2} \le g(t) \quad \forall t \ge 0; \\ \int_0^\infty \left(\int_t^\infty g(s)^2\, ds \right)^{1/2} dt < \infty. \end{cases} \tag{3.60c}$$

As seen before, the gradient map $\mathcal{E}'_F : X_p \to L^2 = H$ satisfies a gradient inequality of LS-type near every critical point in X_p. Hence, direct applications of Theorems 4.1 (ii) to (3.59) under the choice $\mathcal{N} = \mathcal{E}' = \mathcal{E}'_F$ and Theorem 4.4 to (3.60) under the choice $\mathcal{E}' = \mathcal{E}'_F$ yield the following result.

THEOREM 5.1. *(i) Let $u : [0,\infty) \to H$ be a solution of (3.59a,b) such that its orbit $O(u)$ is contained in a compact subset of the Banach space X_p with some $p \ge 2$, $p > N$. Then there exists a $\varphi \in X_p$ such that*

$$\lim_{t\to\infty} \|u(t) - \varphi\|_{W^{2,p}} = 0, \quad \mathcal{E}'_F(\varphi) = 0.$$

(ii) Let $u : [0,\infty) \to H$ be a solution of (3.60a,b) such that

$$\{\dot u(t) : t \ge 0\} \subset X_p, \quad \sup\{\|\dot u(t)\|_{L^2} : t \ge 0\} < \infty$$

and the orbit $O(u) = \{u(t) : t \ge 0\}$ is contained in a compact subset of the Banach space X_p with $p \ge 2$, $p > N$. Then there exists some $\varphi \in X_p$ such that

$$\lim_{t\to\infty} (\|\dot u(t)\|_{L^2} + \|u(t) - \varphi\|_{W^{2,p}}) = 0, \quad \mathcal{E}'_F(\varphi) = 0.$$

Evolutionary systems of the form $\dot u = -\mathcal{E}'_F(u)$ arise from the extensive classical topic known as calculation of variations; see [**27, 112, 122, 43**]. A recently developed applications field in this topic is known as image processing [**12, 5**]. Moreover, many geometric problems can also be put into this category; see [**9, 112**].

To close this section, we also recall the work of L. Simon [**107**]. One of the main interests of L. Simon [**107**] is the application to geometric problems concerning the uniqueness of the tangent cone for minimal surfaces and the uniqueness of the tangent for harmonic maps. In order to prove these uniqueness results, Simon needs the convergence result (his Theorem 1 is related to our Theorem 5.1 (ii)) for solutions to the second-order gradient-like systems $\ddot u - \dot u = \mathcal{E}'_F(u) + g(t,u)$ of elliptic type. (Simon's Theorem 2 is concerned with the convergence of solutions to the first-order gradient-like system (3.59a) with $\mathcal{N} = \mathcal{E}'_F$ and thus is related to our Theorem 5.1 (i) above.)

CHAPTER 4

Applications to semilinear gradient-like systems in Hilbert spaces

Our tasks are twofold. First we will use Theorem 4.2 of Chapter 2 to derive many classical gradient inequalities for semilinear gradient maps. Second, we will apply these gradient inequalities, by virtue of the theorems given in §4 of Chapter 3, to give convergence results for the corresponding semilinear gradient-like systems settled in Hilbert spaces.

1. The generic case

1.1. The setting. Let $(\Omega, \mathcal{B}, \mu)$ be a given measure space with a finite total measure

$$|\Omega| \equiv \mu(\Omega) < \infty. \tag{4.1}$$

Let d be a positive integer. We denote by $L^p(\Omega; \mathbb{R}^d)$ the Banach spaces of vector-valued functions $u = (u_1, ..., u_d)$ of d components that are endowed with the norms

$$\|u\|_{L^p} := \begin{cases} \left(\displaystyle\sum_{j=1}^{d} \int_{\Omega} |u_j|^p \, d\mu \right)^{1/p} & \text{if } \ 1 \le p < \infty, \\ \displaystyle\max_{1 \le j \le d} \|u_j\|_{L^\infty} \quad \text{if } \ p = \infty. \end{cases} \tag{4.2}$$

All spaces $L^p(\Omega; \mathbb{R}^d)$ are Banach lattices (see Nagel et al. [**89**] for this term) where the lattice order $u \le v$ for two vector-valued functions $u = (u_1, ..., u_d)$ and $v = (v_1, ..., v_d)$ is defined as $u_j \le v_j$ for each j. In the sequel, we will denote the spaces $L^p(\Omega; \mathbb{R}^d)$ simply by $L^p(\Omega)$ or by L^p, if no confusion will be caused.

Set

$$\mathcal{H} := L^2(\Omega; \mathbb{R}^d) \tag{4.3}$$

and let

$$A : D(A) \subset L^2(\Omega; \mathbb{R}^d) \to L^2(\Omega; \mathbb{R}^d) \tag{4.4a}$$

be a self-adjoint and positive semidefinite linear operator such that the inverse

$$(1 + A)^{-1} : L^2(\Omega; \mathbb{R}^d) \to L^2(\Omega; \mathbb{R}^d) \quad \text{is compact.} \tag{4.4b}$$

Let

$$B := \sqrt{A} \tag{4.5a}$$

be the square root of A and put

$$\mathcal{E}_F(u) := \frac{1}{2}\|Bu\|_{L^2}^2 + \int_\Omega F(\cdot, u)\, d\mu, \tag{4.5b}$$

where $F : \Omega \times \mathbb{R}^d \to \mathbb{R}$ is a Carathéodory function that is C^2 with respect to the second argument, uniformly to the first argument. In the sequel, we will denote the gradient of F with respect to the u variable by f, i.e.,

$$f \equiv (f_1, ..., f_d) := \nabla_u F. \tag{4.5c}$$

In the above we do not mention the definition domain of the functional $\mathcal{E}_F$. Evidently, the best possible definition domain for $\mathcal{E}_F$ is the one assigned with the operator $B = \sqrt{A}$, i.e., the Hilbert space

$$V := (D(B), \|\cdot\|_B) \equiv (D(\sqrt{A}), \|\cdot\|_{\sqrt{A}}), \tag{4.5d}$$

where the norm on V is the graph norm given by

$$\|u\|_B := (\|u\|_{L^2}^2 + \|Bu\|_{L^2}^2)^{1/2} \quad (u \in V).$$

However, this yields a very strict growth restriction on the function F (we will come back to this point in the next section). Since it is a general situation that no *a priori* growth restriction on the function F should be assumed, the "natural" (but not the best) definition domain of the functional $\mathcal{E}_F$ is the Banach space

$$X := D(B) \cap L^\infty(\Omega; \mathbb{R}^d) = V \cap L^\infty. \tag{4.5e}$$

We see that $\mathcal{E}_F : X \to \mathbb{R}$ is a C^2 functional and

$$\begin{cases} \mathcal{E}_F'(u)v = \langle Bu, Bv\rangle + \langle f(\cdot, u), v\rangle \\ \mathcal{E}_F''(u)vw = \langle Bv, Bw\rangle + \langle F_2(\cdot, u)v, w\rangle \end{cases} \quad (u, v, w \in X), \tag{4.5f}$$

where $f := \nabla_u F$ and $F_2 := \mathrm{Hess}_u F$ is the Hessian matrix of F with respect to the u variable. Moreover, $\langle\cdot,\cdot\rangle$ is the usual scalar product in the Hilbert space $\mathcal{H} = L^2(\Omega; \mathbb{R}^d)$. We have

$$\mathcal{E}_F'(u) = Au + f(x, u) \quad \forall u \in D(A) \cap L^\infty. \tag{4.5g}$$

Our task is to find further conditions so that the abstract theory developed in §4 of Chapter 3 (for the Hilbert spaces case) is applicable to the special choice of the semilinear gradient map $\mathcal{E}' = \mathcal{E}_F'$. More precisely, we want to find further conditions on the linear part A and the function F so that there exist another Hilbert space H and another Banach space $\tilde{X}$ in such a way that the quintuplet $(H, A, \mathcal{E}_F', X, \tilde{X})$ satisfies hypotheses **(H1)**–**(H3)** in §4 of Chapter 2. As seen previously, once this has been done, then gradient inequalities of LS-type for $\mathcal{E}_F'$ as well as convergence results for compact solutions of the corresponding evolution equations are available.

Evolution equations corresponding to the gradient map $\mathcal{E}'_F$ given by (4.5f) are semilinear systems for which there are many tools to determine the existence of solutions (see, e.g., Henry [**63**], Pazy[**94**]). In order to complete our theory, we will also tackle the problem of finding further conditions on the operator A and the function F so that the existence of global and compact solutions for a large class of initial data can be ensured.

Summing up, our tasks are to find further conditions on the operator A and the function F for the following purposes:

- validity of gradient inequality of LS-type
- existence of global and compact solutions

It will be seen below that the conditions for the validity of gradient inequality and those for the global existence of compact solutions are different.

1.2. Sobolev towers. We begin with a preparation.

As before, we assume $A : D(A) \subset \mathcal{H} \to \mathcal{H}$ to be a self-adjoint and positive semidefinite operator in the Hilbert space $\mathcal{H} = L^2(\Omega; \mathbb{R}^d)$ and we let

$$\tilde{A} := 1 + A \quad \text{with the definition domain} \quad D(\tilde{A}) := D(A). \tag{4.6a}$$

$\tilde{A}$ is a self-adjoint and positive definite linear operator. By virtue of the classical spectral theorem, the fractional powers $\tilde{A}^\alpha$ for $\alpha \in \mathbb{R}$ are well-defined linear operators on the following dense subspace of $\mathcal{H}$:

$$D(\tilde{A}^\infty) := \bigcap_{n \geq 1} D(\tilde{A}^n).$$

For each $\alpha \in \mathbb{R}$, we let $\mathcal{H}_\alpha$ be the completion of $D(\tilde{A}^\infty)$ with respect to the norm

$$\|u\|_\alpha := \|\tilde{A}^\alpha u\|_{L^2} \quad (u \in D(A^\infty)). \tag{4.6b}$$

Each $\mathcal{H}_\alpha$ is again a Hilbert space under the inner product $\langle u, v\rangle_\alpha$, given by

$$\langle u, v\rangle_\alpha := \langle \tilde{A}^\alpha u, \tilde{A}^\alpha v\rangle \quad (u, v \in D(\tilde{A}^\infty)),$$

where $\langle \cdot, \cdot\rangle$ is the usual inner product on $\mathcal{H}$. Since for all $u, v \in D(\tilde{A}^\infty)$ we have

$$\langle \tilde{A}u, \tilde{A}v\rangle_\alpha = \langle \tilde{A}^{1+\alpha} u, \tilde{A}^{1+\alpha} v\rangle = \langle u, v\rangle_{1+\alpha},$$

the operator $\tilde{A} : D(\tilde{A}^\infty) \to D(\tilde{A}^\infty)$ extends to an isometric isomorphism

$$\tilde{A}_\alpha : \mathcal{H}_{1+\alpha} \to \mathcal{H}_\alpha. \tag{4.6c}$$

We identify $\mathcal{H}_{1+\alpha}$ as a subspace of $\mathcal{H}_\alpha$. Then

$$\tilde{A}_\alpha : D(\tilde{A}_\alpha) = \mathcal{H}_{1+\alpha} \subset \mathcal{H}_\alpha \to \mathcal{H}_\alpha \tag{4.6d}$$

is a self-adjoint and positive definite operator. We call $\tilde{A}_\alpha$ the *realization* of $\tilde{A}$ in the Hilbert space $\mathcal{H}_\alpha$, and we call the collection of the Hilbert spaces

$$\{\mathcal{H}_\alpha : \alpha \in \mathbb{R}\} \tag{4.6e}$$

the *Sobolev towers* generated by the operator $\tilde{A}$; cf. Engel-Nagel [**34**, p. 126]. The spaces $\mathcal{H}_\alpha$ with $\alpha > 0$ are called *interpolated spaces*, while the spaces $\mathcal{H}_\alpha$ with $\alpha < 0$ are called *extrapolated spaces*. We have

$$\mathcal{H}_\alpha = (D(\tilde{A}^\alpha), \|\cdot\|_{\tilde{A}^\alpha}) \quad \forall \alpha \geq 0. \tag{4.6f}$$

In this way, the Hilbert space V (see (4.5d)) is recovered as

$$V = (D(\sqrt{A}), \|\cdot\|_{\sqrt{A}}) = \mathcal{H}_{1/2}. \tag{4.6g}$$

Moreover, when considering both $\mathcal{H}_\alpha$ and $\mathcal{H}_{-\alpha}$ as Banach spaces, then $\mathcal{H}_{-\alpha}$ can be identified as the Banach space dual $(\mathcal{H}_\alpha)'$ of $\mathcal{H}_\alpha$; i.e., we have

$$\mathcal{H}_{-\alpha} = (\mathcal{H}_\alpha)' \quad (\alpha \geq 0). \tag{4.6h}$$

1.3. Gradient inequalities and their convergence implication. The conditions we need for the validity of gradient inequality read as follows.

• **Compactness of the operator A**

We consider a self-adjoint and positive semidefinite operator

$$A : D(A) \subset L^2(\Omega; \mathbb{R}^d) \to L^2(\Omega; \mathbb{R}^d)$$

such that the inverse

$$(1 + A)^{-1} : L^2(\Omega; \mathbb{R}^d) \to L^2(\Omega; \mathbb{R}^d) \quad \text{is compact.} \tag{4.7a}$$

Since L^∞ is a subspace of L^2 (noting $\mu(\Omega) < \infty$), the restriction $(1 + A)^{-1} : L^\infty \to L^2$ is also compact. Our next condition is to strengthen this compactness: $(1 + A)^{-1}$ keeps the subspace L^∞ invariant and such that the restriction

$$(1 + A)^{-1} : L^\infty(\Omega; \mathbb{R}^d) \to L^\infty(\Omega; \mathbb{R}^d) \quad \text{is compact.} \tag{4.7b}$$

Obviously, the above conditions are satisfied for $A = -\Delta$, where the Laplacian is assigned with one of the classical boundary conditions. In this case, condition (4.7b) is equivalent to the fact that weak solutions of the Laplacian equation $\Delta g - g = h$ with bounded h are bounded again – a certainly well known classical result (see Gilbarg-Trudinger [**45**, Cor. 8.11, p. 186]).

In general, if G is the generator of a symmetric submarkovian semigroup on $(\Omega, \mathcal{B}, \mu)$ with compact resolvent, then $A = -G$ satisfies (4.7); cf. Davies [**30**], Ma-Röckner [**84**].

• **Uniform analyticity of the function F**

$F : \Omega \times \mathbb{R}^d \to \mathbb{R}$ is uniformly analytic with respect to the second argument in the sense that there exists a neighborhood $\Sigma \subset \mathbb{C}^d$ of the real plane $\mathbb{R}^d$ and an extension of F, denoted by $\tilde{F}$, such that $\tilde{F} : \Omega \times \Sigma \to \mathbb{C}$ is locally bounded in $\Omega \times \Sigma$, and $\tilde{F}(x, u)$ is analytic in the u variable and the analyticity is uniform for a.e. $x \in \Omega$. Below we will call this property the *uniform analyticity* with respect to the second argument.

We have the following useful observation. Under the above situation, the composition operator

$$L^\infty(\Omega; \mathbb{R}^d) \ni u \mapsto F(x, u) \in L^\infty(\Omega)$$

is analytic. The proof of this result is very similar to the one for Lemma 5.1 of Chapter 2. Hence we omit the details.

Beginning with the operator A, we construct as in **1.2** the Sobolev towers $\mathcal{H}_\alpha$ ($\alpha \in \mathbb{R}$). We set

$$H := \mathcal{H}_{-1/2} = V', \quad G := \tilde{A}_{-1/2}. \tag{4.8a}$$

Note that here

$$\tilde{A} := 1 + A,$$

$V = (D(B), \|\cdot\|_B) = \mathcal{H}_{1/2}$, and $B = \sqrt{A}$ is the square root of A.

From the above construction, we see that

$$G : V = \mathcal{H}_{1/2} \subset H \to \mathcal{H}_{-1/2} = H \tag{4.8b}$$

is an extension of $\tilde{A}$, and its inverse

$$G^{-1} : H \to H \quad \text{is compact.} \tag{4.8c}$$

Recall also that the space X given by (4.5e) is

$$X = V \cap L^\infty. \tag{4.8d}$$

Since $V = \mathcal{H}_{1/2}$, the following

$$|||u|||_X := \|u\|_{L^\infty} + \|\sqrt{\tilde{A}}u\|_{L^2} \quad (u \in X)$$

defines an equivalent norm on X. Let

$$\tilde{X} := GX \subset H \tag{4.9a}$$

be endowed with the norm

$$\|u\|_{\tilde{X}} := |||G^{-1}u|||_X \quad (u \in \tilde{X}). \tag{4.9b}$$

Then the restriction of G in the subspace $X \subset D(B)$

$$G : X \to \tilde{X} \quad \text{is an isomorphism.} \tag{4.9c}$$

We claim that the embedding

$$L^\infty(\Omega; \mathbb{R}^d) \hookrightarrow \tilde{X} \quad \text{is compact.} \tag{4.9d}$$

To see this, we consider $u \in L^\infty$. By condition (4.7b) we have that $\tilde{A}^{-1}u \in L^\infty$ and thus

$$G(\tilde{A}^{-1}u) = \tilde{A}(\tilde{A}^{-1}u) = u,$$

since G is an extension of $\tilde{A}$. Hence,

$$G^{-1}u = \tilde{A}^{-1}u \in L^\infty \cap D(A) \subset X.$$

It follows that $u \in \tilde{X}$ and

$$\|u\|_{\tilde{X}} = |||\tilde{A}^{-1}u|||_X = \|\left(\sqrt{\tilde{A}}\right)^{-1} u\|_{L^2} + \|\tilde{A}^{-1}u\|_{L^\infty}.$$

Equivalently,

$$\|u\|_{\tilde{X}} = \|\left(\sqrt{\tilde{A}}\right)^{-1} u\|_{L^2} + \|\tilde{A}^{-1}u\|_{L^\infty} \quad \forall u \in L^\infty. \tag{4.9e}$$

Since $\tilde{A}^{-1} : L^2 \to L^2$ is compact by (4.7a), its square root $(\tilde{A}^{-1})^{1/2} = \left(\sqrt{\tilde{A}}\right)^{-1}$ is also a compact operator on L^2. Now let $\{u_n\}$ be a bounded sequence in L^∞. By the compactness of the operators $(\sqrt{\tilde{A}})^{-1} : L^2 \to L^2$ and $\tilde{A}^{-1} : L^\infty \to L^\infty$ (see (4.7b)), we see that $\{u_n\}$ contains a subsequence, say $\{u_{n'}\}$, such that $\{(\sqrt{\tilde{A}})^{-1}u_{n'}\}$ is a Cauchy sequence in L^2 and $\{\tilde{A}^{-1}u_{n'}\}$ is a Cauchy sequence in L^∞. This implies by (4.9e) that $\{u_{n'}\}$ is a Cauchy sequence in the Banach space $\tilde{X}$. Hence, we have proved (4.9d).

From the above proof we also see that there holds

$$X = V \cap L^\infty \subset \tilde{X}.$$

The analyticity of F implies that each component of $f \equiv \nabla_u F$ is again analytic. As seen before, this implies that the composition operator

$$L^\infty(\Omega;\mathbb{R}^d) \ni u \mapsto f(x,u) \in L^\infty(\Omega;\mathbb{R}^d)$$

is analytic. By (4.9d), the restriction of this map in the subspace X, denoted by $\mathcal{K}$, is again an analytic map from X into $\tilde{X}$, which is *compact* in the sense that it maps bounded subsets of X into precompact subsets of $\tilde{X}$. Hence, a result of Krasnoselski [**75**, Theorem 8.1, p. 57] asserts that the linear operators

$$\mathcal{K}'(u) = F_2(x,u) \in L(X,\tilde{X}) \tag{4.10a}$$

are compact for all $u \in X$.

Now we consider the semilinear map $\mathcal{M}_F : X \to \tilde{X}$:

$$\mathcal{M}_F(u) := (G-1)u + f(x,u) \quad (u \in X). \tag{4.10b}$$

Since G is an extension of $\tilde{A} = A + 1$, we have

$$\mathcal{M}_F(u)v = \mathcal{E}'_F(u)v \quad \forall u, v \in X$$

and thus $\mathcal{M}_F$ is an **analytic** gradient map with

$$\mathcal{M}'_F(u) = G + F_2(x,u) \quad \forall u \in X. \tag{4.10c}$$

In the above, $f = \nabla_u F$ and $F_2 = \text{Hess}_u F - 1$. Since G is an isomorphism from X onto $\tilde{X}$ and the product $G^{-1}F_2(x,u) \in L(X,X)$ is compact, the linear operator $\mathcal{M}'_F(u) = G(1 + G^{-1}F_2(x,u))$ is a Fredholm operator of index zero, by the classical Riesz-Schauder theory.

We claim that the quintuplet

$$(H, G, \mathcal{M}_F, X, \tilde{X}) \tag{4.11}$$

constructed as above satisfies hypotheses **(H1)**–**(H3)** in §4 of Chapter 2 with the special property that $\mathcal{M}_F$ is analytic so that Theorem 4.2 of Chapter 2 applies.

Obviously, hypothesis **(H1)** on the pair (H,G) and hypothesis **(H2)** on the pair $(X,\tilde{X})$ have been satisfied. The fact that each operator $\mathcal{M}'_F(u) \in L(X,\tilde{X})$ is a Fredholm operator of index zero has already been established.

Thus, it remains to check hypothesis **(H3)** concerning the existence of the extension $\mathcal{M}_1$ with the required Fredholm property.

For this purpose, we note that, in our case, we have

$$H_G = D(G) = D(B) \subset \mathcal{H} = L^2(\Omega; \mathbb{R}^d)$$

and thus, by (4.8b),

$$G : H_G \to H \quad \text{is an isomorphism.}$$

Thus, by defining

$$\mathcal{M}_1(u)w := Gw + F_2(x,u)w \quad (w \in H_G; u \in X),$$

we obtain for each $u \in X$ a bounded symmetric linear operator $\mathcal{M}_1(u) : H_G \to H$ such that $\mathcal{M}_1(u)w = \mathcal{M}'_F(u)w$ for all $w \in X$. By rewriting each $\mathcal{M}_1(u)$ as

$$\mathcal{M}_1(u) = G(1 + G^{-1}F_2(x,u))$$

and noting that $G^{-1} : H \to H$ is compact (see (4.8c)) and $G : H_G = D(B) \to H$ is an isomorphism, we conclude that each $\mathcal{M}_1(u)$ is a Fredholm operator of index zero. Moreover, for any $u, v \in X$, we have

$$(\mathcal{M}_1(u) - \mathcal{M}_1(v))w = F_2(\cdot, u)w - F_2(\cdot, v)w$$

for all $w \in H_G$. It follows that

$$\|\mathcal{M}_1(u)w - \mathcal{M}_1(v)w\|_H \leq \|F_2(\cdot, u) - F_2(\cdot, v)\|_{L^\infty} \cdot \|w\|_{\mathcal{H}}$$

and thus

$$\|\mathcal{M}_1(u) - \mathcal{M}_1(v)\|_{L(H_G,H)} \leq \|G^{-1}\|_{L(\mathcal{H})} \cdot \|F_2(\cdot, u) - F_2(\cdot, v)\|_{L^\infty}.$$

Consequently, $\|\mathcal{M}_1(u) - \mathcal{M}_1(v)\|_{L(H_G,H)} \to 0$ as $\|u-v\|_X \to 0$ by noting that X is continuously embedded in L^∞ and $F_2(x,\xi)$ is uniformly continuous in ξ. Hence, all required hypotheses **(H1)**–**(H3)** in §4 of Chapter 2 are satisfied.

We restate the conditions used above, for the sake of clarity.

> ♡ $A : D(A) \subset L^2(\Omega; \mathbb{R}^d) \to L^2(\Omega; \mathbb{R}^d)$ is a self-adjoint and positive semidefinite operator such that both inverses
>
> $$(1 + A)^{-1} : L^2 \to L^2, \quad (1 + A)^{-1} : L^\infty \to L^\infty$$
>
> are compact.
>
> ♢ The function $F : \Omega \times \mathbb{R}^d \to \mathbb{R}$ is uniformly analytic with respect to the second argument.

Put

$$V := (D(B), \|\cdot\|_B), \quad X := V \cap L^\infty, \tag{4.12}$$

where $B = \sqrt{A}$ is the square root of A.

As seen before, the above conditions ♡ and ♢ ensure that the quintuplet $(H, G, \mathcal{M}_F, X, \tilde{X})$ satisfies hypotheses **(H1)**–**(H3)** in §4 of Chapter 2 and thus Theorem 4.2 of Chapter 2 applies to the analytic gradient map $\mathcal{E}' = \mathcal{E}'_F$.

It yields that $\mathcal{E}_F'$ satisfies a gradient inequality of LS-type near every critical point: For each critical point $\varphi \in X = V \cap L^\infty$ (i.e., $\mathcal{E}_F'(\varphi) = 0$ in V'), there exist positive constants c, σ, and $\theta \in [1/2, 1)$ such that

$$(*) \qquad c \cdot |\mathcal{E}_F(u) - \mathcal{E}_F(\varphi)|^\theta \le \|Au + f(\cdot, u)\|_{V'}$$

for all $u \in X$ with $\|u - \varphi\|_X < \sigma$. This result can be improved a lot.

THEOREM 1.1. *Under the above conditions ♡ and ◇ there hold the following assertions (i)–(ii):*

(i) Let $\varphi \in V \cap L^\infty$ be a critical point. Then for any $\rho > 0$ there exist positive constants $c \equiv c(\rho), \sigma \equiv \sigma(\rho)$ depending on ρ, and $\theta \in [1/2, 1)$ that is independent from ρ, such that

$$(4.13) \qquad c \cdot |\mathcal{E}_F(u) - \mathcal{E}_F(\varphi)|^\theta \le \|Au + f(\cdot, u)\|_{V'}$$

for all $u \in V \cap L^\infty$ with $|\mathcal{E}_F(u) - \mathcal{E}_F(\varphi)| < \rho$ and $\|u - \varphi\|_{L^\infty} < \sigma$.

(ii) Let K be a bounded subset of $V \cap L^\infty$ that is contained in a compact subset of L^∞. Then there exist a function $\phi \in \mathcal{G}_2$ and a set $U \supset K$ that is open in L^∞ such that

$$(4.14) \qquad \phi(\mathcal{E}_F(u)) \le \|Au + f(\cdot, u)\|_{V'} \quad \forall u \in U.$$

Here $f \equiv \nabla_u F$ is the gradient of F with respect to the u variable.

PROOF. We prove (i) first. As seen before, for the given critical point φ, we can find positive constants c_0, σ_0, and $\theta \in [1/2, 1)$, that are independent from the given constant ρ, such that

$$(4.15a) \qquad c_0 \cdot |\mathcal{E}_F(u) - \mathcal{E}_F(\varphi)|^\theta \le \|Au + f(\cdot, u)\|_{V'}$$

for all $u \in X$ with $\|u - \varphi\|_X < \sigma_0$. We want to show that (4.13) holds true under the exponent θ given in (4.15a). We argue by contradiction; i.e., we assume in contrast that there exists some $\rho > 0$ for which there is a sequence $\{u_n\} \subset X$ such that

$$\|u_n - \varphi\|_\infty \to 0, \quad |\mathcal{E}_F(u_n) - \mathcal{E}_F(\varphi)| < \rho$$

but

$$(4.15b) \qquad (1/n) \cdot |\mathcal{E}_F(u_n) - \mathcal{E}_F(\varphi)|^\theta > \|Au_n + f(\cdot, u_n)\|_{V'}.$$

The convergence $\|u_n - \varphi\|_{L^\infty} \to 0$ as $n \to \infty$ implies that

$$\|f(\cdot, u_n) - f(\cdot, \varphi)\|_{V'} \to 0 \quad \text{as} \quad n \to \infty$$

and thus

$$\|u_n - \varphi\|_V = \|A(u_n - \varphi)\|_{V'} \to 0 \quad \text{as} \quad n \to \infty$$

by (4.15b), combined with the fact that $A\varphi = -f(\cdot, \varphi)$ in V'. Hence,

$$\|u_n - \varphi\|_X = \max\{\|u_n - \varphi\|_V, \|u_n - \varphi\|_{L^\infty}\} \to 0$$

as $n \to \infty$ and thus for sufficiently large n we have $\|u_n - \varphi\|_X < \sigma_0$. It follows that

$$\begin{aligned} c_0 \cdot |\mathcal{E}_F(u_n) - \mathcal{E}_F(\varphi)|^\theta &\le \|Au_n + f(\cdot, u_n)\|_{V'} \quad \text{(by (4.15a))} \\ &< (1/n) \cdot |\mathcal{E}_F(u_n) - \mathcal{E}_F(\varphi)|^\theta \quad \text{(by (4.15b))} \end{aligned}$$

for all sufficiently large n. Consequently, we have that $0 < c_0 < 1/n$ for all sufficiently large n, and this is impossible.

To prove (ii), we note that the boundedness of K in $V \cap L^\infty$ implies that the range $\mathcal{E}_F(K)$ is bounded, say $\mathcal{E}_F(K) \subset [-M, M]$. Then $|\mathcal{E}_F(u) - \mathcal{E}_F(v)| \le 2M$ for all $u, v \in K$. We apply assertion (i) to each $u \in K$. It yields for each $\varphi \in K$ (either a critical point or not) the existence of three positive constants $c(\varphi) \in (0, 1/2), \sigma(\varphi)$, and $\theta(\varphi) \in [1/2, 1)$, depending only on u, such that

$$c(\varphi) \cdot |\mathcal{E}_F(u) - \mathcal{E}_F(\varphi)|^{\theta(\varphi)} \le \|Au + f(\cdot, u)\|_{V'} \tag{4.16}$$

for all $u \in K$, $\|u - \varphi\|_{L^\infty} < \sigma(\varphi)$.

For each $\varphi \in K$, we define the neighborhood

$$U(\varphi) := \{u \in L^\infty : \|u - \varphi\|_{L^\infty} < \sigma(\varphi)\}.$$

Then $\{U(\varphi) : \varphi \in K\}$ forms an open cover of the precompact subset $K \subset L^\infty$. Hence, there exist finitely many points $\{\varphi_i : i = 1, 2, ..., N\} \subset K$ such that $K \subset \bigcup_{i=1}^N U(\varphi_i) =: U$. It follows from (4.16) that

$$\|Au + f(\cdot, u)\|_{V'} \ge \phi(\mathcal{E}_F(u)) \quad \forall u \in U,$$

where

$$\phi(s) := \min_{1 \le i \le N} c(\varphi_i) \cdot \min\{|s - \mathcal{E}_F(\varphi_i)|^{\theta(\phi_i)}, |s - \mathcal{E}_F(\varphi_i)|^{1/2}\}$$

for all $s \in \mathbb{R}$. By the remarks following Definition 2.1 of Chapter 2, we have that $\phi \in \mathcal{G}_2$. Clearly, the set U is open in L^∞. We are done. □

As a consequence of the validity of gradient inequalities given by Theorem 1.1, we have the following convergence result.

THEOREM 1.2. *Under the above conditions ♡ and ◇, Theorem 4.2 (ii), Theorem 4.4 as well as Theorem 4.5 of Chapter 3 are applicable to the case where*

$$V := D(\sqrt{A}); \quad (H, X, \mathcal{E}') := (L^2, V \cap L^\infty, \mathcal{E}'_F).$$

More exactly, under the above choice of the spaces V, H and X, and the map $\mathcal{E}'$, there hold the following assertions (i)–(ii):

(i) If $u : [0, \infty) \to V \cap L^\infty$ is a solution of equation (3.28a) with $\mathcal{E}' = \mathcal{E}'_F$ such that its orbit $O(u)$ is precompact in $V \cap L^\infty$, and the time-dependent perturbation g satisfies the smallness condition (3.28c′), then there exists $\varphi \in V \cap L^\infty$ such that

$$\lim_{t \to \infty} \|u(t) - \varphi\|_{V \cap L^\infty} = 0, \quad \mathcal{E}'_F(\varphi) = 0.$$

(ii) Assume that $u : [0, \infty) \to V \cap L^\infty$ *is a solution of (3.32a) with* $\mathcal{E}' = \mathcal{E}'_F$ *and with either sign choice. If*

$$\{\dot{u}(t) : t \geq 0\} \subset V \cap L^\infty, \quad \sup\{\|\dot{u}(t)\|_{L^2} : t \geq 0\} < \infty,$$

the orbit $O(u)$ *is precompact in* $V \cap L^\infty$*, and there holds respectively the boundedness condition (3.33b) for the positive sign choice and condition (3.38b) with* $V = D(\sqrt{A})$ *for the negative sign choice, then there exists some* $\varphi \in V \cap L^\infty$ *such that*

$$\lim_{t \to \infty} (\|\dot{u}(t)\|_{L^2} + \|u(t) - \varphi\|_{V \cap L^\infty}) = 0, \quad \mathcal{E}'_F(\varphi) = 0.$$

PROOF. There is nothing to do for case (i).

For case (ii) with the positive sign, we need only to check the condition (3.33a); i.e., there holds

$$(*) \qquad \langle \mathcal{E}''_F(u(t))\dot{u}(t), \dot{u}(t)\rangle_{L^2} \geq -c_1 \|\dot{u}(t)\|^2_{L^2} \quad \forall t \geq 0$$

with some constant $c_1 > 0$. To see this, we consider $w := u(t) \in O(u)$ and let $v := \dot{u}(t)$. Using (4.5f) we have that

$$(**) \qquad \langle \mathcal{E}''_F(w)v, v\rangle = \|Bv\|^2 + \langle F_2(\cdot, w)v, v\rangle \geq -|\langle F_2(\cdot, w)v, v\rangle|,$$

where $B = \sqrt{A}$ is the square root of A. Now the boundedness of the orbit $O(u)$ combined with the uniform analyticity of the map F_2 implies that there exists some constant $c_1 > 0$ such that $|\langle F_2(\cdot, w)v_1, v_1\rangle| \leq c_1 \|v_1\|^2$ for all $v_1 \in L^2$. Hence, by $(**)$ we see that $(*)$ is satisfied by this constant c_1. The desired convergence result then follows from an application of Theorem 4.4 of Chapter 3.

For case (ii) with the negative sign we need to show that both conditions (3.38a) and (3.38c) are satisfied with the Hilbert space $V = (D(B), \|\cdot\|_B)$. By Theorem 1.1 (ii) we see that the gradient map $\mathcal{E}' : X \to V'$ satisfies a gradient inequality of LS-type near every cluster point of the precompact orbit $O(u) \subset X$. Therefore, condition (3.38a) is satisfied under the choice $V = D(B)$.

To see that the second condition (3.38c) is also satisfied by this choice, we note that the duality map $K : V' \to V$ is given by $K := A^{-1}$. Hence, for $w \in X = V \cap L^\infty$ and $v \in D(A)$, we have that

$$K \circ \mathcal{E}''_F(w)v = A^{-1}(Av + f(\cdot, w)v) = v + A^{-1} f(\cdot, w)v.$$

From this we infer that each operator $K \circ \mathcal{E}''_F(w)$ $(w \in X)$ admits an extension to a bounded linear operator on L^2 and $K \circ \mathcal{E}''_F : X \to L(L^2)$ maps bounded sets into bounded sets. In particular, we see that the norms $\|K \circ \mathcal{E}''_F(w)\|_{L(L^2)}$ $(w \in O(u))$ are uniformly bounded, since the orbit $O(u)$ is bounded in $X = V \cap L^\infty$. Thus, condition (3.38c) is satisfied as well. The desired convergence result then follows from an application of Theorem 4.5 of Chapter 3. □

1.4. Existence of global solutions and stationary solutions. We follow the unpublished work [**67**] of the author. We are working in the Hilbert space

$$\mathcal{H} = L^2(\Omega; \mathbb{R}^d).$$

In the following, for a scalar real-valued function v on Ω, we use $v^{\pm}$ to denote the positive/negative part of the function, i.e., $v^+(x) := \max\{v(x), 0\}$ and $v^-(x) := \max\{-v(x), 0\}$ for all $x \in \Omega$. This notion is extended to vector-valued functions in a direct way; i.e., if $v = (v_1, ..., v_n)$ is a vector-valued and real-valued function on Ω, then the positive/negative part $v^{\pm}$ is defined by $v^{\pm} := (v_1^{\pm}, ..., v_n^{\pm})$.

Our setup is as follows.

♠ **Submarkovian property and compactness of the generator** A

• **Submarkovian property** The linear operator

$$A : D(A) \subset L^2(\Omega; \mathbb{R}^d) \to L^2(\Omega; \mathbb{R}^d)$$

is a diagonal operator matrix of the form

$$A = \operatorname{diag}(A_1, A_2, ..., A_d), \quad D(A) = D(A_1) \times D(A_2) \times \cdots \times D(A_d),$$

where each $A_j : D(A_j) \subset L^2(\Omega) \to L^2(\Omega)$ is a self-adjoint and positive semi-definite operator that generates a *symmetric submarkovian* C_0 *semigroup* $\{e^{-tA_j} : t \geq 0\}$ on $L^2(\Omega)$ (see Davies [**30**] and Ma-Röckner [**84**] for a general theory of such semigroups). Equivalently, each operator e^{-tA_j} is positivity preserving (i.e., $e^{-tA_j}u \geq 0$ for all $u \geq 0$) and extends to a contraction on every $L^p(\Omega)$, $1 \leq p \leq \infty$, i.e.,

$$\|e^{-tA_j}u\|_{L^p} \leq \|u\|_{L^p} \quad \forall u \in L^2 \cap L^p;\ 1 \leq p \leq \infty. \tag{4.17a}$$

By the Beurling-Deny characterization theorem (see Davies [**30**], Ma-Röckner [**84**]), this is equivalent to the following single condition:

$$\sum_{j=1}^{d} \langle A_j u_j, (u_j - 1)^+ \rangle_{L^2} \geq 0 \quad \forall (u_1, u_2, ..., u_d) \in D(A). \tag{4.17a$'$}$$

Clearly, (4.17a′) implies and is equivalent to the following one:

$$\sum_{j=1}^{d} \langle A_j u_j, (u_j - r_j)^+ \rangle_{L^2} \geq 0 \quad \forall u = (u_1, u_2, ..., u_d) \in D(A) \tag{4.17a$''$}$$

for each nonnegative constant vector $(r_1, r_2, ..., r_d) \geq 0$.

In the sequel, we will call an operator A with the above mentioned submarkovian property a **submarkovian generator**.

• **Compactness** The embedding

$$X = D(\sqrt{A}) \cap L^\infty(\Omega; \mathbb{R}^d) \hookrightarrow L^2(\Omega; \mathbb{R}^d) \quad \text{is compact.} \tag{4.17b}$$

Under the above conditions for A, for $0 < \alpha \leq 1$ and for $2 \leq p < \infty$, we let

$$Y(\alpha, p) := D(A^\alpha \restriction L^p) \cap L^\infty = \{u \in L^\infty : A^\alpha u \in L^p\}$$

be endowed with the norm

$$\|u\|_{Y(\alpha,p)} := \max\{\|u\|_{L^\infty}, \|A^\alpha u\|_{L^p}\}. \tag{4.17c}$$

The subspace $X = V \cap L^\infty$ is recovered as $Y(\frac{1}{2}, 2)$, i.e., we have $Y(\frac{1}{2}, 2) = D(\sqrt{A}) \cap L^\infty$.

♣ **Smoothness and growth condition on the function** F

We do not need the analyticity of F. For the smoothness we assume only that $F : \Omega \times \mathbb{R}^d \to \mathbb{R}$ is uniformly C^{2-} with respect to the second argument; i.e., F is uniformly C^1 with respect to the second argument and its gradient $f \equiv \nabla_u F$ is locally Lipschitz in the sense that for any $R > 0$ there exists some $L \equiv L(R) > 0$ such that

$$|f(\cdot, u) - f(\cdot, v)| \le L \cdot |u - v| \quad \forall u, v \in \mathbb{R}^d, |u| + |v| \le R. \tag{4.18a}$$

For the growth condition, we assume first that

$$f(\cdot, 0) \in L^\infty(\Omega; \mathbb{R}^d) \tag{4.18b}$$

so that f is locally uniformly bounded. Moreover, we assume that there exist three constant vectors $\tau_+ \equiv (\tau_+^1, \tau_+^2, ..., \tau_+^d) \ge 0, \tau_- \equiv (\tau_-^1, \tau_-^2, ..., \tau_-^d) \le 0$, and $M \equiv (M_1, M_2, ..., M_d) \ge 0$ such that for all boundary points $u \equiv (u_1, ..., u_d) \in \partial Q_\tau$ of the rectangle $Q_\tau := \{u \in \mathbb{R}^d : \tau_- \le u \le \tau_+\}$, there hold the inequalities:

$$\begin{cases} f_j(\cdot, u) \ge M_j & \text{if} \quad u_j = \tau_+^j; \\ f_j(\cdot, u) \le -M_j & \text{if} \quad u_j = \tau_-^j; \end{cases} \qquad j = 1, 2, ..., d. \tag{4.18c}$$

Recall that $(f_1, f_2, ..., f_d)$ are the components of the map f.

It is worth noting that, for the scalar case $d = 1$, condition (4.18c) is equivalent to the following very transparent one:

$$f(\cdot, \tau_+) \ge M, \quad f(\cdot, \tau_-) \le -M. \tag{4.18c$'$}$$

with the constants $\tau_+ \ge 0, \tau_- \le 0$, and $M \ge 0$.

Before going to the next step, we derive one consequence from the above condition (4.17a). For each $p \in [2, \infty)$ and $\alpha \in [0, 1)$, there exists a constant $c = c(\alpha, p) > 0$, depending on α and p, such that

$$\|A^\alpha e^{-tA} u\|_{L^p} \le c \cdot t^{-\alpha} \cdot \|u\|_{L^p} \quad \forall u \in L^\infty, t > 0. \tag{4.19}$$

This is a consequence of a deep result of V. A. Liskevich and Yu. A. Semenov [**80**] (see also Huang [**66**], Nagel-Voigt [**90**], and Stein [**111**]), which says that for each $p \in (1, \infty)$ the contractive semigroup $\{e^{-tA} : t \ge 0\}$ on $L^p(\Omega; \mathbb{R}^d)$ extends to an analytic semigroup $\{T(z)\}$ of contractions on $L^p(\Omega; \mathbb{R}^d)$, defined in the sector $\{z \in \mathbb{C} : |\arg(z)| < \arccos|1 - 2/p|\}$. Hence, (4.19) follows from the theory of analytic semigroups; cf. Henry [**63**, Theorem 1.4.3].

Below we will turn our attention to the existence of global and compact solutions as well as stationary solutions to the following Cauchy problem:

$$\begin{cases} \dot{u} = -\mathcal{E}_F'(u) + g(t, u), \quad u(0) = u_0, \\ \mathcal{E}_F'(u) \equiv Au + f(x, u), \end{cases} \tag{4.20a}$$

where the perturbation

$$g : \mathbb{R}_+ \times L^\infty(\Omega; \mathbb{R}^d) \to L^\infty(\Omega; \mathbb{R}^d)$$

is locally Hölder continuous in time, locally Lipschitz continuous for the second argument, and uniformly bounded:

$$-M \le g(t,u) \le M \quad \forall (t,u) \in \mathbb{R}_+ \times L^\infty, \tag{4.20b}$$

where $M \ge 0$ is the constant vector involved in condition (4.18c). Moreover, we assume that the perturbation $g(t,u)$ satisfies the following smallness condition: There exists some function $g \in L^2(\mathbb{R}_+)$ such that

$$\|g(t,u)\|_{L^2} \le g(t) \quad \forall (t,u) \in \mathbb{R}_+ \times L^\infty. \tag{4.20c}$$

We begin with one more preparation.

Claim: Under conditions (4.18a,b,c) there exists for each given $R > 0$ a constant $c_1 \equiv c_1(R) \ge 0$ such that

$$\begin{aligned}&(u-\tau_+)^+ \bullet (M - f(\cdot,u)) + (\tau_- - u)^+ \bullet (M + f(\cdot,u)) \\ &\le c_1 \cdot \left(|(u-\tau_+)^+|^2 + |(\tau_- - u)^+|^2\right)\end{aligned} \tag{4.18c$''$}$$

for all $u \equiv (u_1, ..., u_d) \in \mathbb{R}^d$ with $|u| \le R$. In the sequel, we denote the Euclidean norm and scalar product by $|\cdot|$ and $\bullet$, respectively.

We remark that (4.18c$''$) implies in turn the growth condition (4.18c).

To prove the above claim, we assume that $R > 0$ has been chosen so large that the rectangle $Q_\tau = \{u \in \mathbb{R}^d : \tau_- \le u \le \tau_+\}$ is contained in the closed ball $\{|u| \le R\}$. Using the local Lipschitz continuity of f, we can find a constant $L \equiv L(R) > 0$ such that (4.18a) holds true.

Consider a point $u \equiv (u_1, ..., u_d) \in \mathbb{R}^d$ with $|u| \le R$ and let

$$S := (u-\tau_+)^+ \bullet (M - f(\cdot,u)) + (\tau_- - u)^+ \bullet (M + f(\cdot,u)).$$

If u is already located in Q_τ, then we have that $(\tau_- - u)^+ = 0 = (u - \tau_+)^+$ and thus $S = 0$. Hence, it suffices to treat the case where u is not located in Q_τ. For this case, we may assume, without loss of generality, that there is some $k \le d$ such that

$$\begin{cases} u_j \notin [\tau_-^j, \tau_+^j] & \text{for} \quad j = 1, ..., k; \\ u_j \in [\tau_-^j, \tau_+^j] & \text{for} \quad j > k. \end{cases}$$

Note that the second line will be voided if $k = d$. For each j, $1 \le j \le k$, we choose $\tilde{u}_j \in \{\tau_-^j, \tau_+^j\}$ so that

$$|\tilde{u}_j - u_j| = \min\{|u_j - \tau_-^j|, |u_j - \tau_+^j|\}.$$

Equivalently, we have that $\tilde{u}_j := \tau_+^j$ if $u_j > \tau_+^j$, and $\tilde{u}_j := \tau_-^j$ if $u_j < \tau_-^j$.

Let $\tilde{u} \in \partial Q_\tau$ be the vector obtained by replacing the first k components of u by the new components $\tilde{u}_j$ and keeping the other components of u invariant, i.e.,

$$\tilde{u} := (\tilde{u}_1, ..., \tilde{u}_k, u_{k+1}, ..., u_d).$$

Applying condition (4.18c) to the boundary point $\tilde{u}$, we find that

$$\begin{cases} f_j(\cdot, \tilde{u}) \geq M_j & \text{if} \quad u_j > \tau_+^j; \\ f_j(\cdot, \tilde{u}) \leq -M_j & \text{if} \quad u_j < \tau_-^j. \end{cases}$$

It follows that

$$(\tau_-^j - u_j)^+(M_j + f_j(\cdot, \tilde{u})) + (u_j - \tau_+^j)^+(M_j - f_j(\cdot, \tilde{u})) \leq 0$$

for all $j = 1, 2, ..., k$ and thus

$$S1 := \sum_{j=1}^{k} [(\tau_-^j - u_j)^+(M_j + f_j(\cdot, \tilde{u}) + (u_j - \tau_+^j)^+(M_j - f_j(\cdot, \tilde{u})))] \leq 0.$$

Note that

$$(\tau_-^j - u_j)^+ = 0 = (u_j - \tau_+^j)^+ = 0 \quad \forall j > k.$$

Therefore, we have that

$$S = S1 + S2,$$

where

$$S2 := \sum_{j=1}^{k} ((\tau_-^j - u_j)^+ - (u_j - \tau_+^j)^+)(f_j(\cdot, u) - f_j(\cdot, \tilde{u})).$$

Since

$$S2 \leq |f(\cdot, u) - f(\cdot, \tilde{u})| \cdot \sum_{j=1}^{k} ((\tau_-^j - u_j)^+ + (u_j - \tau_+^j)^+)$$

by the Lipschitz continuity condition (4.18a), we have that

$$S2 \leq L \cdot |u - \tilde{u}| \cdot \sum_{j=1}^{k} ((\tau_-^j - u_j)^+ + (u_j - \tau_+^j)^+).$$

On the other hand, we have by the definition of $\tilde{u}$ that

$$|u - \tilde{u}|^2 = \sum_{j=1}^{k} |u_j - \tilde{u}_j|^2 \leq |(\tau_- - u)^+|^2 + |(u - \tau_+)^+|^2.$$

Moreover, it is easily seen that

$$\begin{aligned} &\sum_{j=1}^{k} ((\tau_-^j - u_j)^+ + (u_j - \tau_+^j)^+) \\ &\leq \sqrt{k} \cdot \Big(\sum_{j=1}^{k} \left(|(\tau_-^j - u_j)^+|^2 + |(u_j - \tau_+^j)^+|^2\right)\Big)^{1/2} \\ &\leq d \cdot (|(\tau_- - u)^+|^2 + |(u - \tau_+)^+|^2)^{1/2}. \end{aligned}$$

Taking these estimates together, we obtain finally that

$$S \leq S2 \leq (dL) \cdot (|(\tau_- - u)^+|^2 + |(u - \tau_+)^+|^2).$$

Therefore, the desired inequality (4.18c″) holds true under the choice of the constant $c_1 := dL$.

Our first result implies particularly that the rectangle Q_τ is an invariant region for the gradient-like system (4.20), provided that the nonlinearity f satisfies conditions (4.18a,b,c).

LEMMA 1.3. **(Comparison principle)**
Assume conditions ♠ and ♣.
(i) **(Valid for the vectorial case)** *Let*

$$u \in C([0,T);L^2) \cap C((0,T);D(A)) \cap C^1((0,T);L^2)$$

be a classical solution of (4.20a,b) such that

$$\sup\{\|u(t)\|_{L^\infty} : t \in [0,T)\} < \infty. \tag{4.21a}$$

If the initial value $u_0 = u(0) \in L^\infty$ is such that

$$\tau_- \le u_0 \le \tau_+, \tag{4.21b}$$

then there holds

$$\tau_- \le u(t) \le \tau_+ \quad \forall t \in [0,T). \tag{4.21c}$$

Here $\tau_+ \ge 0$ and $\tau_- \le 0$ are the constant vectors involved in the growth condition (4.18c).

(ii) **(Valid only for the scalar case $d = 1$)** *Assume $d = 1$. Let*

$$u, v \in C([0,T);L^2) \cap C^1((0,T);L^2) \cap C((0,T);D(B))$$

be two maps such that

$$\sup\{\|u(t)\|_{L^\infty} + \|v(t)\|_{L^\infty} : t \in [0,T)\} < \infty, \tag{4.21d}$$

$$u(0) \le v(0) \tag{4.21e}$$

and there holds for almost every $t \in (0,T)$ the inequality

$$\dot u(t) + \mathcal{E}_F'(u(t)) - g(t,u(t)) \le \dot v(t) + \mathcal{E}_F'(v(t)) - g(t,v(t)) \tag{4.21f}$$

in the distributional sense; i.e., there holds for all $0 \le \psi \in D(B)$ that

$$\begin{aligned} &\langle \psi, \dot u(t) + \mathcal{E}_F'(u(t)) - g(t,u(t))\rangle \\ &\le \langle \psi, \dot v(t) + \mathcal{E}_F'(v(t)) - g(t,v(t))\rangle. \end{aligned} \tag{4.21f$'$}$$

Then one has that

$$u(t) \le v(t) \quad \forall t \in [0,T). \tag{4.21g}$$

PROOF. **Proof of (i):** Consider the function

$$N(t) := \frac{1}{2}\int_\Omega (|(u(t)-\tau_+)^+|^2 + |(\tau_- - u(t))^+|^2)\, d\mu$$

for all $t \in [0,T)$. We have

$$N(0) = 0,$$

since $u(0) - \tau_+ = u_0 - \tau_+ \le 0$ and $\tau_- - u(0) = \tau_- - u_0 \le 0$ by (4.21b).

Using the representation $v^+ = (v + |v|)/2$ for a function $v : \Omega \to \mathbb{R}$, we have that

$$(v^+)^2 = \frac{1}{2}v^2 + \frac{1}{2}v|v|$$

and thus

$$\frac{\partial}{\partial t}(v^+)^2 = v\frac{\partial}{\partial t}v + \frac{1}{2}|v|\frac{\partial}{\partial t}v + \frac{1}{2}v\frac{\partial}{\partial t}|v|.$$

It follows from the formula (see Gilbarg-Trudinger [**45**, Lemma 7.6, p. 152], or Ziemer [**126**, Cor. 2.1.8, p. 47])

$$\frac{\partial}{\partial t}|v| = \begin{cases} \frac{\partial}{\partial t}v & \text{if} \quad v > 0 \\ 0 & \text{if} \quad v = 0 \\ -\frac{\partial}{\partial t}v & \text{if} \quad v < 0 \end{cases}$$

that $v\frac{\partial}{\partial t}|v| = |v|\frac{\partial}{\partial t}v$ and thus

$$\frac{\partial}{\partial t}(v^+)^2 = (v + |v|)\frac{\partial}{\partial t}v = 2v^+\frac{\partial}{\partial t}v.$$

Applying this result to the functions $v := \tau_+^j - u_j(t)$ and $v := u_j(t) - \tau_-^j$ $(t \in [0,T); j = 1, 2, ..., d)$, we obtain for a.e. $t \in (0,T)$ that

$$\frac{dN(t)}{dt} = \int_\Omega ((u - \tau_+)^+ - (\tau_- - u)^+) \bullet \dot{u}\, d\mu.$$

By substituting $\dot{u} = -Au - f(x,u) + g(t,u)$ and using $Au = (A_1u_1, ..., A_du_d)$, we obtain that

$$\begin{aligned} dN(t)/dt \quad &= \quad -\sum_{j=1}^{d}\langle (u_j - \tau_+^j)^+ - (\tau_-^j - u_j)^+, A_ju_j\rangle_{L^2} \\ &\quad + \int_\Omega ((u - \tau_+)^+ - (\tau_- - u)^+) \bullet (-f(x,u) + g(t,u))\, d\mu. \end{aligned}$$

Note that $\sum_{j=1}^d \langle A_ju_j, (u_j - \tau_+^j)^+\rangle_{L^2} \geq 0$ and

$$\sum_{j=1}^{d}\langle A_ju_j, -(\tau_-^j - u_j)^+\rangle_{L^2} = \sum_{j=1}^{d}\langle A_j(-u_j), ((-u_j) - (-\tau_-^j))^+\rangle_{L^2} \geq 0$$

by (4.17a″), since $\pm u_j \in D(A_j)$ and $\pm\tau_\pm \geq 0$. Moreover, $\|g(t,u)\|_{L^\infty} \leq M$ by (4.20b), since $u \in L^\infty$ by (4.21a). Therefore,

$$\begin{aligned} \frac{dN(t)}{dt} \quad &\leq \quad \int_\Omega (u - \tau_+)^+ \bullet (M - f(x,u))\, d\mu \\ &\quad + \int_\Omega (\tau_- - u)^+ \bullet (M + f(x,u))\, d\mu. \end{aligned}$$

As prepared before (see (4.18c″)), condition (4.18c) combined with condition (4.18a) implies that the integrand is not larger than

$$c_1 \cdot [|(u - \tau_+)^+|^2 + |(\tau_- - u)^+|^2]$$

with a constant $c_1 \geq 0$, depending on the value $\sup\{\|u(t)\|_{L^\infty} : t \in [0,T)\} < \infty$. Therefore,

$$\frac{dN(t)}{dt} \leq c_1 \int_{\tilde{\Omega}} [|(u-\tau_+)^+|^2 + |(\tau_- - u)^+|^2]\, d\mu = (2c_1)N(t),$$

which is equivalent to

$$\frac{d}{dt}(\exp(-2c_1 t)N(t)) \leq 0.$$

By integration we find that

$$N(t) \leq \exp(2c_1 t)N(0) \quad \forall t \in [0,T).$$

But $N(0) = 0$, we must have that $N(t) = 0$ for all $t < T$. Equivalently, we have that $(u(t) - \tau_+)^+ = 0 = (\tau_- - u(t))^+$ and thus $\tau_- \leq u(t) \leq \tau_+$ for all $t \in [0,T)$.

Proof of (ii): The proof is completely similar to that for (i). At present we have $d = 1$. We consider the difference

$$w(t) := u(t) - v(t) \quad (t \in [0,T))$$

and the norm function

$$N_1(t) := \frac{1}{2}\int_\Omega [w(t)^+]^2\, d\mu \quad (t \in [0,T)).$$

We want to show that $N_1(t) = 0$ for all $t \in [0,T)$.

We need an extra result stating that the domain $D(B)$ is closed under the lattice operations $\vee, \wedge$, and $|\cdot|$. This follows from the general Theorem 4.12 of Ma-Röckner [**84**, p. 36] (see the latter Lemma 1.6). In particular, we find that the positive parts $w(t)^+ = w(t) \vee 0$ are in $D(B)$, since $w(t) \in D(B)$ for all $t \in (0,T)$.

We have

$$N_1(0) = 0,$$

since $w(0) = u(0) - v(0) \leq 0$ by (4.21e). Consider the map

$$h(t,x,w) := -f(x,w) + g(t,w) \quad ((t,x,w) \in \mathbb{R}_+ \times \Omega \times \mathbb{R}).$$

It follows from the local Lipschitz continuity of both maps f and g that h is locally Lipschitz continuous, uniformly for $x \in \Omega$ and $t \geq 0$. Thus, under the boundedness condition (4.21d) we can find a constant $c_2 > 0$ such that

$$(*) \qquad |h(t,x,u(t)) - h(t,x,v(t))| \leq c_2 \cdot |u(t) - v(t)|$$

for all $t \in [0,T); x \in \Omega$. We have

$$\frac{d}{dt}N_1(t) = \int_\Omega w(t)^+ \dot{w}(t)\, d\mu.$$

By condition (4.21f) we have that

$$\dot{w}(t) \leq -Aw(t) + [h(t,x,u(t)) - h(t,x,v(t))]$$

for almost every $t \in (0,T)$, in the distributional sense. This yields, by noting the estimate in $(*)$ and that $w(t)^+ \in D(B) \geq 0$, that

$$\frac{d}{dt}N_1(t) \leq -\langle Bw(t)^+, Bw(t)\rangle_{L^2 \times L^2} + c_2 \cdot \int_\Omega w(t)^+ |w(t)|\, d\mu.$$

Using the inequality (4.17a″) with $r_1 = 0$ for the submarkovian generator $A = A_1$, we see that $-\langle Bw(t)^+, Bw(t)\rangle_{L^2 \times L^2} \leq 0$. Moreover, since $d = 1$, each $w(t)$ is a scalar function and thus there holds $w(t)^+|w(t)| = [w(t)^+]^2$. Hence,

$$\frac{d}{dt}N_1(t) \leq c_2 \cdot \int_\Omega [w(t)^+]^2\, d\mu = (2c_2) \cdot N_1(t) \quad \text{a.e. } t \in (0,T).$$

By integration, we find that $0 \leq N_1(t) \leq \exp(2c_2 t)N_1(0) \leq 0$ and thus $N_1(t) = 0$ for all $t \in [0,T)$. This completes the proof. □

Remarks to Lemma 1.3 (Comparison Principle). (a) It is worth noting that the compactness requirement for the operator A stated in ♠ (see (4.17b)) is not used in the above proof of Lemma 1.3. In fact, what we have used for the operator A is only its submarkovian property. Moreover, the property that the nonlinearity $f = \nabla_u F$ is a gradient map in ♣ is also not needed in the above proof of Lemma 1.3. Hence, the above Comparison Principle (Lemma 1.3) holds true for any submarkovian symmetric generator A and any locally Lipschitzian nonlinearity f that satisfies conditions (4.18a,b,c).

(b) In the classical case where $A = -\Delta$ is the Laplacian accompanying the classical boundary conditions, the comparison principle in Lemma 3.1 (ii) (for the scalar case $d = 1$) is well known. The usual way to prove this is to use the classical *Maximum Principle* (see Gilbarg-Trudinger [**45**, Theorem 2.2, p. 15]). However, no analogy to the Maximum Principle for the Laplacian operators is known for a general submarkovian generator A. Note that our above proof for Lemma 1.3 (ii) also remains valid for the case where the measure $|\Omega|$ is infinite.

(c) For the scalar case $d = 1$ we can also define the notion **lower/upper solutions** (also called **sub-/supersolution**) as in the classical literature (cf., e.g., Smoller [**110**, Chapter 10]) as follows. We call a map

$$u \in C([0,T); L^2) \cap C((0,T); D(B)) \cap C^1((0,T); L^2)$$

a lower (resp., upper) solution of (4.20a) if $\dot{u}(t) + \mathcal{E}_F'(u(t) - g(t, u(t)) \leq 0$ for almost every $t \in [0,T)$ (resp., the reverse inequalities hold), in the distributional sense. Using this notion, the comparison principle in Lemma 1.3 (ii) implies the following: Assume A is a submarkovian generator. If u is a bounded lower solution of (4.20a) in the interval $[0,T)$ and v is a bounded upper solution of (4.20a) in the same interval $[0,T)$ such that $v(0) \geq u(0)$, then there holds $v(t) \geq u(t)$ for all $t \in [0,T)$. We point out that in our present work we do not use the notion of lower/upper solutions, since it is not needed.

Our main global existence result is as follows.

THEOREM 1.4. *Assume conditions ♠ and ♣. Moreover, assume the conditions (4.20b,c) for the perturbation* $g(t,u)$.

Then, for every initial value $u_0 \in L^\infty(\Omega;\mathbb{R}^d)$ *with*

$$\tau_- \le u_0 \le \tau_+, \tag{4.22a}$$

the Cauchy problem (4.20a) has a unique global classical solution

$$u \in C([0,\infty);L^2) \cap C((0,\infty);D(A)) \cap C^1((0,\infty);L^2) \tag{4.22b}$$

such that

$$\tau_- \le u(t) \le \tau_+ \quad \forall t \ge 0 \tag{4.22c}$$

and for which there holds the energy equality

$$\frac{d}{dt}\mathcal{E}(u(t)) = \langle \mathcal{E}_F'(u(t)), \dot{u}(t)\rangle_{L^2} \quad \forall t > 0. \tag{4.22d}$$

Moreover, the section $\{u(t) : t \ge 1\}$ *is bounded in every Banach space* $Y(\alpha,p)$ *in the sense that*

$$\sup_{t\ge 1} \|u(t)\|_{Y(\alpha,p)} \le C(\alpha,p), \tag{4.22e}$$

where the constant $C(\alpha,p)$ *depends only on the pair* $(\alpha,p) \in (0,1)\times[2,\infty)$, *and is independent of the solution* u.

Particularly, the section $\{u(t) : t \ge 1\}$ *is precompact in* $L^2(\Omega;\mathbb{R}^d)$ *and its omega-limit set* $\omega(u_0)$ *in* $L^2(\Omega;\mathbb{R}^d)$ *is contained in the set of critical points of* $\mathcal{E}_F$. *More precisely, there holds*

$$\omega(u_0) \subset \mathcal{S}_\tau \cap D(A), \tag{4.22f}$$

where

$$\mathcal{S}_\tau := \{\varphi \in X : \mathcal{E}_F'(\varphi) = 0,\ \tau_- \le \varphi \le \tau_+\}. \tag{4.22g}$$

PROOF. Our proof method for the existence of global solutions is a "bootstrap" method.

Let $\sigma := 2(|\tau_+| + |\tau_-|) + 1$ and consider the retraction map $r : \mathbb{R}^d \to \mathbb{R}^d$ given by

$$r(u) := \begin{cases} u & \text{if} \quad |u| \le \sigma \\ \frac{\sigma}{|u|}u & \text{if} \quad |u| > \sigma. \end{cases} \tag{4.23a}$$

It is a geometric fact that r is a global Lipschitz map with Lipschitz constant 1. (To check the inequality $|r(u) - r(v)| \le |u - v|$, one uses isometric transformations to see that it suffices to check the inequality for the case where both u, v are points in the two-dimensional plane and u locates itself in the positive half real-axis.) Combining this with the local Lipschitz continuity of both f and g, we see that the truncated map $H : \mathbb{R}_+ \times L^2 \to L^2$, given by

$$H(t,u) := -f(\cdot, r(u)) + g(t, r(u)) \quad (t \ge 0, u \in L^2), \tag{4.23b}$$

satisfies **Assumption (F)** in Pazy [**94**, p. 196] (stating that H is locally Hölder continuous in time and locally Lipschitz continuous in u). Moreover, using the locally uniform boundedness of f and the uniform boundedness of g (see (4.20b)), we find that the range of H is contained in a bounded subset of L^∞; i.e., there exists some constant $c_2 > 0$ such that

$$\|H(t,u)\|_{L^\infty} \le c_2 \quad \forall t \ge 0; u \in L^2. \tag{4.23c}$$

This implies in particular that the range of H is contained in a bounded subset of L^2, since the embedding $L^\infty \hookrightarrow L^2$ is continuous. Therefore, Theorem 3.3 in Pazy [**94**, p. 199] applies and yields that the following Cauchy problem:

$$\dot{u} = -Au + H(t,u), \quad u(0) = u_0 \tag{4.24a}$$

with the given initial value $u_0 \in L^\infty$ satisfying (4.22a) has a unique global classical solution u satisfying (4.22b).

We need to show that this solution of (4.24a) satisfies (4.22c). Once this has been established, we find from the definition of the retraction map r that $r(u(t)) = u(t)$ and thus $H(t,u(t)) = -f(\cdot,u(t)) + g(t,u(t))$ for all $t \ge 0$, which yields that u is a solution of the original equation (4.20a).

To show (4.22c), we rewrite Eq. (4.24a) in its integral form:

$$u(t) = e^{-tA}u_0 + \int_0^t e^{-sA}H(t-s,u(t-s))\,ds \quad (t \ge 0). \tag{4.24b}$$

Since each operator e^{-tA} is a contraction on L^∞ (see (4.17a)), we have that

$$\|e^{-tA}u_0\|_{L^\infty} \le \|u_0\|_{L^\infty} \le \max\{|\tau_+|,|\tau_-|\} \le \sigma - 1$$

and

$$\begin{aligned}
&\left\|\int_0^t e^{-sA}H(t-s,u(t-s))\,ds\right\|_{L^\infty} \\
&\le \int_0^t \|e^{-sA}H(t-s,u(t-s))\|_{L^\infty}\,ds \le c_2 t
\end{aligned}$$

by (4.23c). Hence,

$$\|u(t)\|_{L^\infty} \le \|u_0\|_{L^\infty} + c_2 t \le \sigma + (c_2 t - 1) \quad \forall t \ge 0.$$

This implies that $\|u(t)\|_{L^\infty} \le \sigma$ for all $t \le T_0$, where $T_0 := 1/c_2 > 0$ is a constant. Consequently, we have that $r(u(t)) = u(t)$ and hence $H(t,u(t)) = -f(\cdot,u(t)) + g(t,u(t))$ for all $t \in [0,T_0]$. In other words, u solves the original Eq. (4.20a) in the time interval $[0,T_0]$ and hence, by Lemma 1.3 we find that $\tau_- \le u(t) \le \tau_+$ for all $t \in [0,T_0]$. Continuing this procedure with the new starting time point T_0 at which one has $\tau_- \le u(T_0) \le \tau_+$, a similar approach yields that u satisfies the inequality $\tau_- \le u(t) \le \tau_+$ for all $t \in [T_0, 2T_0]$; i.e., there holds the inequality $\tau_- \le u(t) \le \tau_+$ for all $t \le 2T_0$. A continuing argument then yields the desired inequality (4.22c) for all $t \ge 0$.

The energy equality (4.22d) follows directly from the fact that the global solution u is a classical solution for all $t > 0$.

We're left to prove the boundedness of the section $K := \{u(t) : t \geq 1\}$ in each $Y(\alpha, p)$ with a bound that is independent of the solution u. To this end, we note that (4.22c) implies that K is a bounded subset in L^∞ with the bound $\|\tau_-\|_{L^\infty} + \|\tau_+\|_{L^\infty}$.

To find the boundedness property of K under the norms $\|A^\alpha \cdot \|_{L^p}$, we note that u solves the equation

$$\dot{u} = -(A+1)u + h(t), \tag{4.25a}$$

where

$$h(t) := u(t) - f(\cdot, u(t)) + g(t, u(t)) \quad (t \geq 0). \tag{4.25b}$$

Let

$$c_3 := \operatorname{ess\,sup}\{|f(x,s)| : x \in \Omega; s \in \mathbb{R}^d, |s| \leq \sigma\} < \infty.$$

Then we have

$$\|h(t)\|_{L^\infty} \leq c_4 \quad \forall t \geq 0, \tag{4.25c}$$

where $c_4 := \sigma + c_3 + M$. We rewrite Eq. (4.22a) in its integral form:

$$u(t) = e^{-tA-t}u_0 + \int_0^t e^{-sA-s}h(t-s)\,ds \quad (t \geq 0). \tag{4.25d}$$

It follows for all $\alpha \in (0,1)$ that

$$A^\alpha u(t) = A^\alpha e^{-tA-t}u_0 + \int_0^t A^\alpha e^{-sA-s}h(t-s)\,ds \quad (t > 0).$$

This implies, by (4.19), for all $p \in [2, \infty)$ that

$$\|A^\alpha u(t)\|_{L^p} \leq c \cdot t^{-\alpha}\|u_0\|_{L^p} + \int_0^t c \cdot s^{-\alpha}e^{-s}\|h(t-s)\|_{L^p}\,ds.$$

Note that $\|u_0\|_{L^p} \leq \sigma(1 + |\Omega|)$ and by (4.25c) that $\|h(s)\|_{L^p} \leq c_4(1 + |\Omega|)$ for all $s \geq 0$. We claim from above that the integral on the right-hand side is uniformly bounded for all $t \geq 0$ and thus we claim finally that

$$\sup_{t \geq 1}\|A^\alpha u(t)\|_{L^p} \leq c(1 + |\Omega|)(\sigma + c_4\Gamma(1-\alpha)) < \infty \tag{4.25e}$$

for each pair $(\alpha, p) \in (0,1) \times [2, \infty)$. In the above, we have replaced the integral $\int_0^\infty s^{-\alpha}e^{-s}\,ds$ by the Gamma-function $\Gamma(1-\alpha)$. The number on the right-hand side of (4.25e) is independent of the solution u. Therefore, we have proved the estimate in (4.22e).

In particular, K is a bounded subset of $Y(\frac{1}{2}, 2) = X$, and thus is pre-compact in L^2, since the embedding $X \hookrightarrow L^2$ is compact by assumption (see (4.17b)).

To finish the proof, we need to show the inclusion $\omega(u_0) \subset \mathcal{S}_\tau \cap D(A)$. To this end, we let $\varphi \in \omega(u_0)$. Then there exists an unbounded and increasing sequence $\{t_n\} \subset [1, \infty)$ such that

$$u(t_n) \to \varphi \quad \text{in} \quad L^2 \quad \text{as} \quad n \to \infty. \tag{4.26a}$$

Clearly, we have $\tau_- \le \varphi \le \tau_+$. We use the energy equality (4.22d) and take integration over the time interval $[1, \infty)$. It yields that

$$\int_1^\infty \langle \mathcal{E}_F'(u(t)), -\dot{u}(t)\rangle \, dt < \infty.$$

Using $\dot{u} = -\mathcal{E}_F'(u) + g(t,u)$, we rewrite the integrand as

$$\langle \mathcal{E}_F'(u), -\dot{u}\rangle = \|\mathcal{E}_F'(u)\|_{L^2}^2 - \langle \mathcal{E}_F'(u), g(t,u)\rangle.$$

By the estimate

$$|\langle \mathcal{E}_F'(u), g(t,u)\rangle| \le (\|\mathcal{E}_F'(u)\|_{L^2}^2 + \|g(t,u)\|_{L^2}^2)/2$$

combined with the estimate $\|g(t,u(t))\|_{L^2} \le g(t)$ in (4.20c), we find that

$$\langle \mathcal{E}_F'(u), -\dot{u}\rangle \ge (\|\mathcal{E}_F'(u)\|_{L^2}^2 - g(t)^2)/2.$$

Since $g \in L^2(\mathbb{R}_+)$, this implies that

$$\text{(4.26b)} \qquad \int_1^\infty \|\mathcal{E}_F'(u(t))\|_{L^2}^2 \, dt < \infty \quad \text{as well as} \quad \int_1^\infty \|\dot{u}(t)\|_{L^2}^2 \, dt < \infty.$$

On the other hand, using the equality

$$u(t_n + s) - u(t_n) = \int_{t_n}^{t_n+s} \dot{u}(\tau)\, d\tau \quad (s > 0),$$

we have that

$$\begin{aligned} \|u(t_n + s) - u(t_n)\|_{L^2} &\le \int_{t_n}^{t_n+s} \|\dot{u}(\tau)\|_{L^2}\, d\tau \\ &\le \sqrt{s} \cdot \left(\int_{t_n}^{t_n+s} \|\dot{u}(\tau)\|_{L^2}^2\, d\tau \right)^{1/2} \end{aligned}$$

for all $s \ge 0$ and thus

$$\text{(4.26c)} \qquad \sup_{0 \le s \le 1} \|u(t_n + s) - u(t_n)\|_{L^2} \le a_n \quad (n \in \mathbb{N}),$$

where

$$a_n := \left(\int_{t_n}^{t_n+1} \|\dot{u}(\tau)\|_{L^2}^2\, d\tau \right)^{1/2} \to 0 \quad \text{as} \quad n \to \infty$$

by (4.26b).

Let $v \in D(A)$. Then we have

$$\langle v, \dot{u}(t)\rangle_{L^2} = -\langle v, \mathcal{E}_F'(u(t))\rangle_{L^2} + \langle v, g(t, u(t))\rangle$$

for all $t > 0$. Note that

$$\int_{t_n}^{t_n+1} \langle v, \dot{u}(t)\rangle_{L^2}\, dt = v \bullet (u(t_n + 1) - u(t_n))$$

and

$$\langle v, \mathcal{E}_F'(u(t))\rangle_{L^2} = \langle Av, u(t)\rangle_{L^2} + \langle v, f(\cdot, u(t))\rangle_{L^2}.$$

We obtain from above that

$$\int_{t_n}^{t_n+1} \langle Av, u(t)\rangle_{L^2}\,dt + \int_{t_n}^{t_n+1} \langle v, f(\cdot, u(t))\rangle_{L^2}\,dt$$
$$= \int_\Omega v \bullet (u(t_n) - u(t_n+1))\,d\mu + \int_{t_n}^{t_n+1} \langle v, g(t, u(t))\rangle\,dt.$$

By (4.26c) combining with (4.26a), we have that

$$\int_{t_n}^{t_n+1} \langle Av, u(t)\rangle_{L^2}\,dt \to \langle Av, \varphi\rangle_{L^2} \quad \text{and}$$

$$\int_\Omega v \bullet (u(t_n) - u(t_n+1))\,d\mu \to 0$$

as $n \to \infty$. Moreover, since $\tau_- \le u(t) \le \tau_+$ for all $t > 0$ and f is locally uniformly bounded, by virtue of the Lebesgue Dominated Convergence Theorem we also have that

$$\int_{t_n}^{t_n+1} \langle v, f(\cdot, u(t))\rangle_{L^2}\,dt \to \langle v, f(\cdot, \varphi)\rangle_{L^2}$$

as $n \to \infty$. Finally, using the estimate

$$|\langle v, g(t, u(t))\rangle| \le \|v\|_{L^2} \cdot g(t),$$

we have that

$$\left|\int_{t_n}^{t_n+1} \langle v, g(t, u(t))\rangle\,dt\right| \le \int_{t_n}^{t_n+1} |\langle v, g(t, u(t))\rangle|\,dt$$
$$\le \|v\|_{L^2} \int_{t_n}^{t_n+1} g(t)\,dt \le \|v\|_{L^2} \left(\int_{t_n}^{t_n+1} g(t)^2\,dt\right)^{1/2} \to 0$$

as $n \to \infty$, since $g \in L^2(\mathbb{R}_+)$. Consequently, we have proved that

$$\langle Av, \varphi\rangle_{L^2} + \langle v, f(\cdot, \varphi)\rangle_{L^2} = 0$$

for all $v \in D(A)$. Equivalently,

$$\langle (1 + A)v, \varphi\rangle_{L^2} + \langle v, f(\cdot, \varphi) - \varphi\rangle_{L^2} = 0$$

for all $v \in D(A)$. But the operator $1 + A$ is invertible; this implies that

$$\varphi + (1 + A)^{-1}(f(\cdot, \varphi) - \varphi) = 0.$$

Note that $f(\cdot, \varphi) - \varphi \in L^\infty \subset L^2$. We claim that

$$\varphi = (1 + A)^{-1}(\varphi - f(\cdot, \varphi)) \in D(A).$$

Thus, $\varphi \in X \cap D(A)$ and $\mathcal{E}_F'(\varphi) = 0$. Hence the desired inclusion $\omega(u_0) \subset \mathcal{S}_\tau \cap D(A)$ follows. □

Before proceeding to other convergence results, we give two results about the solutions of the semilinear equation $Au + f(x,u) = 0$, i.e., the critical points of the functional $\mathcal{E}_F$. For this purpose, we set

$$X_\tau := \{u \in X : \tau_- \le u \le \tau_+\} \tag{4.27a}$$

and consider the following local infimum

$$\alpha_F(\tau) := \inf\{\mathcal{E}_F(u) : u \in X_\tau\} > -\infty. \tag{4.27b}$$

A critical point $\varphi \in \mathcal{S}_\tau = \mathcal{S} \cap X_\tau$ is called a **local ground state** of $\mathcal{E}_F$ in X_τ if $\mathcal{E}_F(\varphi) = \alpha_F(\tau)$, i.e., if

$$\mathcal{E}_F(\varphi) \le \mathcal{E}_F(u) \quad \forall u \in X_\tau. \tag{4.27c}$$

Note that the growth condition (4.18c) (cf. (4.18c$'$) for the scalar case) does not in general imply that the functional $\mathcal{E}_F : X \to \mathbb{R}$ itself is bounded below. Hence, the following result that asserts the attainability of α_F is certainly interesting. More importantly, note that for the validity of this result we do not need higher smoothness of the function F. In fact, the local Lipschitz continuity of its gradient $f = \nabla_u F$ suffices. Certainly, the growth condition (4.18c) does play an important rôle here.

PROPOSITION 1.5. **(Existence of local/global ground states)** *Under conditions ♠ and ♣ for the submarkovian operator A and the nonlinearity f, the following assertions (i)–(ii) are true:*

(i) The subset of equilibria

$$\mathcal{S}_\tau = \{\varphi \in X : \mathcal{E}'_F(\varphi) = 0, \tau_- \le \varphi \le \tau_+\}$$

is a nonempty compact subset of $D(A)$, and is contained and bounded in every Banach space $Y(\alpha,p)$ with $(\alpha,p) \in (0,1) \times [2,\infty)$.

(ii) There exists some $\varphi \in \mathcal{S}_\tau$ such that $\mathcal{E}_F(\varphi) = \alpha_F(\tau)$, i.e.,

$$\mathcal{E}_F(\varphi) \le \mathcal{E}_F(u) \quad \forall u \in X, \tau_- \le u \le \tau_+.$$

Furthermore, the local ground state φ is an absolute ground state; i.e., there holds

$$\mathcal{E}_F(\varphi) \le \mathcal{E}_F(u) \quad \forall u \in X,$$

provided that the function F satisfies the following growth condition:

$$F(\cdot, (u \wedge \tau_+) \vee \tau_-) \le F(\cdot, u) \quad \forall u \in \mathbb{R}^d. \tag{4.28}$$

Moreover, for the scalar case $d = 1$ there holds the following extra assertion:

(iii) **(Valid for the scalar case)** *Assume $d = 1$. Let u_{min} (resp., u_{max}) be the solution of the gradient system $\dot{u} = -\mathcal{E}'_F(u)$ with the minimal (resp., maximal) initial value $u_{min}(0) = \tau_-$ (resp., $u_{max}(0) = \tau_+$). Then there exist $\psi_{min} \in \mathcal{S}_\tau$ and $\psi_{max} \in \mathcal{S}_\tau$ such that*

$$\lim_{t\to\infty} \Big(\|u_{min}(t) - \psi_{min}\|_{L^2} + \|u_{max}(t) - \psi_{max}\|_{L^2} \Big) = 0,$$

and

$$\psi_{min} \le \psi \le \psi_{max} \quad \forall \psi \in \mathcal{S}_\tau.$$

As a consequence, if $\mathcal{S}_\tau$ is not totally ordered, then it contains at least four critical points of $\mathcal{E}_F$.

The assertions in Proposition 1.5 (iii) are taken from the unpublished work [**67**] of the author himself.

Before proving the above results, we cite an additional result about submarkovian generators that has been proved in Ma-Röckner [**84**]. Note that, in [**84**], the property that A is a submarkovian generator in our sense is equivalent to the property that the form

$$D(B) \ni u \mapsto \|Bu\|_{L^2}^2$$

is a *Dirichlet form.* Recall that [**84**, p. 35] a function $T : \mathbb{R} \to \mathbb{R}$ with $T(0) = 0$ is called a *normal contraction* if

$$|T(s) - T(t)| \leq |s - t| \quad \forall s, t \in \mathbb{R}.$$

Typical normal contractions are the usual lattice operations $|\cdot|$, $\cdot \wedge \tau$ with constants $\tau \geq 0$, and $\cdot \vee \tau$ with constants $\tau \leq 0$. The following is a consequence of the general result of Ma-Röckner [**84**, Theorem 4.12, p. 36].

LEMMA 1.6. *Let A be the diagonal submarkovian generator given in ♠, and let $B := \sqrt{A}$ be the square root. Assume that $T = (T_1, T_2, ..., T_d) : \mathbb{R}^d \to \mathbb{R}^d$ is a diagonal map, where each $T_j : \mathbb{R} \to \mathbb{R}$ is a normal contraction. Then, for any $u \in D(B)$, we have $T(u) \in D(B)$ and there holds*

$$\|BT(u)\|_{L^2} \leq \|Bu\|_{L^2} \quad \forall u \in D(B).$$

In particular, this implies that $D(B)$ is a lattice.

Proof of Proposition 1.5: To show (i), we note that the set $\mathcal{S}_\tau$ is nonempty, contained in a bounded subset of $D(A)$, and is bounded in every Banach space $Y(\alpha, p)$, as seen in the proof of Theorem 1.4.

We show the compactness of $\mathcal{S}_\tau$ in $D(A)$. To this end, we consider a sequence $\{\varphi\}_{n\in\mathbb{N}} \subset \mathcal{S}_\tau \subset X \cap D(A)$. Then, we have that

$$(*) \qquad A\varphi_n = -f(x, \varphi_n) \quad \forall n \in \mathbb{N}.$$

The bounded sequence $\{\varphi_n\} \subset X$ is precompact in L^2, since the embedding $X \hookrightarrow L^2$ is compact by assumption. Therefore, by passing to a subsequence if necessary, we can assume that there exists some $\varphi \in L^2$ such that $\|\varphi_n - \varphi\|_{L^2} \to 0$ as $n \to \infty$. Note that the sequence $\{f(\cdot, \varphi_n)\}$ is uniformly bounded and $|\Omega| < \infty$. Hence, by the Lebesgue Dominated Convergence Theorem we have that

$$\lim_{n\to\infty} \|f(\cdot, \varphi_n) - f(\cdot, \varphi)\|_{L^2} = 0.$$

This implies by $(*)$ that $\|A\varphi_n + f(\cdot, \varphi)\|_{L^2} \to 0$ as $n \to \infty$. Thus, by the closedness of the generator A, we find that $\varphi \in D(A)$ and $A\varphi + f(x, \varphi) = 0$. Hence, $\varphi \in \mathcal{S}_\tau$ and there holds the convergence $\|\varphi_n - \varphi\|_{D(A)} \to 0$ as $n \to \infty$. This is exactly the assertion in (i).

To show (ii), we choose a sequence $\{\varphi_n\} \subset \mathcal{S}_\tau$ such that

$$(**) \qquad \mathcal{E}_F(\varphi_n) \to \alpha_F(\tau) \quad \text{as} \quad n \to \infty.$$

By (i), the sequence $\{\varphi_n\}$ is precompact in $D(A)$. Therefore, by passing to a subsequence if necessary, we may assume that there exists some $\varphi \in \mathcal{S}_\tau$ such that

$$(* * *) \qquad \|\varphi_n - \varphi\|_{D(A)} \to 0 \quad \text{as} \quad n \to \infty.$$

Since the sequence $\{f(\cdot, \varphi_n)\}$ is uniformly bounded and $|\Omega| < \infty$, we have by the Lebesgue Dominated Convergence Theorem again that

$$\lim_{n\to\infty} \int_\Omega F(x, \varphi_n)\, d\mu = \int_\Omega F(x, \varphi)\, d\mu.$$

Combining this with $(**)$ and $(* * *)$, we obtain that $\mathcal{E}_F(\varphi) = \alpha_F(\tau)$.

To show that the local ground state φ is an absolute ground state under condition (4.28), we apply Lemma 1.6 to the diagonal (normal contraction) map $Tu := (u \wedge \tau_+) \vee \tau_-$ $(u \in \mathbb{R}^d)$ to find for all $u \in X$ that

$$w := (u \wedge \tau_+) \vee \tau_- \in X, \quad \|Bw\|_{L^2} \le \|Bu\|_{L^2}.$$

Combining this with condition (4.28), we have that

$$\begin{aligned} \mathcal{E}_F(w) &= \frac{1}{2}\|Bw\|_{L^2}^2 + \int_\Omega F(x, (u(x) \wedge \tau_+) \vee \tau_-)\, dx \\ &\le \frac{1}{2}\|Bu\|_{L^2}^2 + \int_\Omega F(x, u(x))\, dx = \mathcal{E}_F(u). \end{aligned}$$

But $\mathcal{E}_F(w) \ge \mathcal{E}_F(\varphi)$, since $w \in X_\tau$. Therefore, we have shown that φ is an absolute ground state.

We show (iii) for the convergence of the minimal solution u_{min}. The proof for the maximal solution is completely similar and thus will be omitted.

For each fixed $s > 0$ we see that $w(t) := u_{min}(t + s)$ $(t \ge 0)$ is also a solution of the gradient system $\dot{u} = -\mathcal{E}'_F(u)$, with the initial value $w(0) = u(s) \ge \tau = u_{min}(0)$. Hence, by the comparison principle (Lemma 1.3) we have that $w(t) \ge u_{min}(t)$ for all $t \ge 0$. Thus, there holds

$$(* * **) \qquad u_{min}(s) \le u_{min}(t) \quad \forall 0 \le s \le t.$$

By Theorem 1.4, the section $\{u_{min}(t) : t \ge 1\}$ is precompact in L^2, and the omega-limit set $\omega(u_{min})$ is contained in $\mathcal{S}_\tau$.

We need to show that $\omega(u_{min})$ is a singleton set. To this end, let $\psi_1, \psi_2 \in \omega(u_{min})$. We choose two unbounded sequences $\{t_n^1\}$ and $\{t_n^2\}$ such that

$$t_n^1 \le t_n^2 \le t_{n+1}^1 \quad \forall n \in \mathbb{N},$$

and

$$\lim_{n\to\infty} (\|u_{min}(t_n^1) - \psi_1\|_{L^2} + \|u_{min}(t_n^2) - \psi_2\|_{L^2}) = 0.$$

Using $(* * **)$, we have that

$$u_{min}(t_n^1) \le u_{min}(t_n^2) \le u_{min}(t_{n+1}^1)$$

for all $n \in \mathbb{N}$. Hence, by taking limits we obtain that $\psi_1 \le \psi_2 \le \psi_1$, i.e., $\psi_1 = \psi_2$. This establishes the desired convergence $u_{min}(t) \to \psi_{min} \in \mathcal{S}_\tau$ as $t \to \infty$.

Next, for any critical point $\psi \in \mathcal{S}_\tau$, we have that $u_{min}(t) \leq \psi \leq u_{max}(t)$ for all $t \geq 0$, again by the comparison principle. Hence, there holds $\psi_{min} \leq \psi \leq \psi_{max}$.

Finally, if $\mathcal{S}_\tau$ is not totally ordered, then there are elements $\psi_3, \psi_4 \in \mathcal{S}_\tau$ such that

$$\psi_{min} \leq \psi_3 \wedge \psi_4 \lneqq \psi_3, \psi_4 \lneqq \psi_3 \vee \psi_4 \leq \psi_{max}.$$

From this we see that the four elements $\psi_{min}, \psi_3, \psi_4$ and ψ_{max} in $\mathcal{S}_\tau$ are distinct. This completes the proof. □

In the next subsection we will deal with the convergence of solutions of the gradient-like system (4.20a), by virtue of gradient inequalities given in Theorem 1.1 of this chapter. Note that there we need a compactness for the operator A given in ♡, namely, that both inverses $(1 + A)^{-1} : L^2 \to L^2$ and $(1 + A)^{-1} : L^\infty \to L^\infty$ are compact. The compactness of the operator $(1 + A)^{-1}$ in the L^2 norm is in general very easy to establish. However, the compactness in the L^∞ norm is sometimes a little difficult to obtain. Because of this reason, we give a criterion below. Recall that the Banach spaces $Y(\alpha, p)$ given by (4.17c) are

$$Y(\alpha, p) = D(A^\alpha \restriction L^p) \cap L^\infty(\Omega; \mathbb{R}^d)$$

with pairs $(\alpha, p) \in (0, 1) \times [2, \infty)$.

PROPOSITION 1.7. *Assume that A satisfies conditions in ♠. Then the inverse*

$$(1 + A)^{-1} : L^\infty(\Omega; \mathbb{R}^d) \to L^\infty(\Omega; \mathbb{R}^d)$$

exists and continuously maps L^∞ into every Banach space $Y(\alpha, p)$.

It follows that the inverse $(1 + A)^{-1} : L^\infty \to L^\infty$ is compact, provided that there exists some pair $(\beta, q) \in (0, 1) \times [2, \infty)$ such that the embedding

$$Y(\beta, q) \hookrightarrow L^\infty(\Omega; \mathbb{R}^d) \tag{4.29}$$

is compact.

PROOF. Fix $h = (h_1, ..., h_d) \in L^\infty(\Omega; \mathbb{R}^d)$ and let $\tau := \|h\|_\infty$. We apply Proposition 1.5 to the following special choice of the F function: $F(x, u) := |u|^2/2 - h(x) \bullet u$. It follows that $f := \nabla_u \tilde{F}$ has the form $f(x, u) = u - h(x)$, or equivalently, $f \equiv (f_1, ..., f_d)$ have components given by

$$f_j(x, u) = u_j - h_j(x) \quad (j = 1, 2, ..., d).$$

Take $\tau_+ := (\tau, \tau, ..., \tau)$, $\tau_- := -\tau_+$ and let $Q_\tau := \{u \in \mathbb{R}^d : \tau_- \leq u \leq \tau_+\}$. Then, for any point $u \in \partial Q_\tau$, we have the following assertions: If $u_j = \tau$, then

$$f_j(x, u) = u_j - h_j(x) \geq 0;$$

if $u_j = -\tau$, then

$$f_j(x, u) = u_j - h(x) \leq 0.$$

Therefore, the function F given as above satisfies the growth condition in (4.18c) with $M = 0$. Hence, by Proposition 1.5 we see that the set

$$K_h := \{u \in D(A) : Au + u - h = 0, \quad \tau_- \leq u \leq \tau_+\}$$

is nonempty, and is contained in every Banach space $Y(\alpha,p)$. We show that K_h contains only one element. In fact, if $u, v \in K_h$, then we have that $Aw + w = 0$ with $w := u - v \in D(A)$. It follows that

$$0 = \langle (Aw + w), w\rangle = \|w\|_{L^2}^2 + \langle Aw, w\rangle.$$

But $\langle Aw, w\rangle \geq 0$, since the operator A is positive semidefinite. Hence, we have shown that, for each $h \in L^\infty$, the inverse $(1+A)^{-1}h \in L^\infty$ exists, and $(1+A)^{-1}h \in Y(\alpha,p)$ for every pair $(\alpha,p) \in (0,1)\times[2,p)$. This implies by the classical Resonance Theorem that the linear operators

$$(1+A)^{-1} : L^\infty \to Y(\alpha,p)$$

are bounded, giving the desired result. □

1.5. Convergence result and stability of ground states. We will present below the convergence results for the case where the nonlinearity f is analytic. Moreover, we will show that the local ground state is uniformly stable in an appropriate sense; this will be defined below.

Our conditions for the convergence of global solutions to (4.20) are obtained by comparing conditions (♡, ♢) for the validity of gradient inequality given in **1.3** and conditions (♠, ♣) for the existence of compact global solutions given in **1.4**. For clarity, we collect these conditions together as follows:

Conditions on A:

- $A = \operatorname{diag}(A_1, A_2, ..., A_d) : D(A) \subset L^2(\Omega;\mathbb{R}^d) \to L^2(\Omega;\mathbb{R}^d)$ is a diagonal, positive semidefinite and symmetric submarkovian generator such that the inverse

$$(1+A)^{-1} : L^2 \to L^2$$

is compact.

- There exist $\beta \in (0,1)$ and $q \in [2,\infty)$ such that the embedding

$$D(A^\beta \restriction L^q) \cap L^\infty(\Omega;\mathbb{R}^d) \hookrightarrow L^\infty(\Omega;\mathbb{R}^d)$$

is compact.

Conditions on F:

Set $f \equiv (f_1, ..., f_d) := \nabla_u F$. Then $f(\cdot,0) \in L^\infty$, and there exist three constant vectors $\tau_+ = (\tau_+^1, \tau_+^2, ..., \tau_+^d) \geq 0, \tau_- = (\tau_-^1, \tau_-^2, ..., \tau_-^d) \leq 0$ and $M = (M_1, M_2, ..., M_d) \geq 0$ such that for all boundary points $u \equiv (u_1, ..., u_d) \in \partial Q_\tau$ of the rectangle $Q_\tau := \{u \in \mathbb{R}^d : \tau_- \leq u \leq \tau_+\}$, there hold the inequalities:

$$\begin{cases} f_j(\cdot,u) \geq M_j & \text{if} \quad u_j = \tau_+^j; \\ f_j(\cdot,u) \leq -M_j & \text{if} \quad u_j = \tau_-^j; \end{cases} \qquad j = 1, 2, ..., d.$$

Let us make some remarks to these conditions. Note that the above given conditions for the function F are just the growth conditions in ♣ of **1.4**. For the operator A, we note that the compactness condition (4.17b) in ♠ of **1.4** for the operator A has been included in ♡ of **1.3**. Therefore, for the operator A, we need conditions in ♡ of **1.3** plus the submarkovian property in ♠ of **1.4**, as given above. Moreover, we see by Proposition 1.7 that the above imposed compactness of the embedding $Y(\beta, q) \hookrightarrow L^\infty$ implies the compactness of the inverse $(1+A)^{-1} : L^\infty \to L^\infty$, which is required in **1.3** for the validity of gradient inequalities.

Finally, for the perturbation, we impose the following smallness condition: $g : \mathbb{R}_+ \times L^\infty(\Omega; \mathbb{R}^d) \to L^\infty(\Omega; \mathbb{R}^d)$ is a locally Lipschitz map satisfying

$$\text{(4.30a)} \quad \begin{cases} -M \le g(t,u) \le M, \ \|g(t,u)\|_{L^2} \le g(t) \quad \forall (t,u) \in \mathbb{R}_+ \times L^\infty; \\ \displaystyle\int_0^\infty \rho(t) g(t)^2 \, dt + \int_0^\infty \frac{1}{\rho(t)} \, dt < \infty \\ \text{with some increasing positive function} \quad \rho. \end{cases}$$

In (4.30a), M is the same constant vector involved in the above growth condition for the nonlinearity $f = \nabla_u F$. Note that in (4.30a) we have imposed two kinds of smallness conditions on the perturbation g. The first one, $-M \le g(t,u) \le M$, corresponding to condition (4.20b), ensures the applicability of Theorem 1.4 while the second one, $\|g(t,u)\|_{L^2} \le g(t)$, where the norms are evaluated in the smaller L^2 norm, corresponds to the smallness condition (3.28c′) involved in Theorem 4.2 (ii) of Chapter 3, which is required for the expected convergence result.

Put

$$\text{(4.30b)} \qquad X := D(\sqrt{A}) \cap L^\infty \quad \text{and} \quad X_\tau := \{u \in X : \tau_- \le u \le \tau_+\},$$

where $\tau_+ \ge 0$ and $\tau_- \le 0$ are constant vectors involved in the above growth condition on the nonlinearity F.

We have the following convergence theorem for analytic gradient-like systems.

THEOREM 1.8. *Let the above conditions given in the boxes for the operator A and the nonlinearity F be satisfied. Assume furthermore that the function $F : \Omega \times \mathbb{R}^d \to \mathbb{R}$ is uniformly analytic with respect to the second argument, and the perturbation g satisfies condition (4.30a).*

Then, for each initial value $u_0 \in L^\infty(\Omega; \mathbb{R}^d)$ with

$$\text{(4.30c)} \qquad \tau_- \le u_0 \le \tau_+,$$

the following Cauchy problem

$$\text{(4.30d)} \qquad \begin{cases} \dot{u} = -\mathcal{E}_F'(u) + g(t,u), & u(0) = u_0 \\ \mathcal{E}_F'(u) \equiv Au + f(x,u), & f \equiv \nabla_u F \end{cases}$$

has a unique, global and classical solution

$$\text{(4.30e)} \qquad u \in C([0,\infty); L^2) \cap C((0,\infty); D(A)) \cap C^1((0,\infty); L^2)$$

for which there exists an equilibrium $\varphi \in X_\tau$ *(i.e.,* $\mathcal{E}_F'(\varphi) = 0$ *and* $\tau_- \le \varphi \le \tau_+$*) such that*

$$\lim_{t\to\infty} \|u(t) - \varphi\|_X = 0. \tag{4.30f}$$

Moreover, the subset of equilibria

$$\mathcal{S}_\tau := \{\varphi \in X_\tau : \mathcal{E}_F'(\varphi) = 0\} \subset D(A) \cap L^\infty \tag{4.30g}$$

is compact in X.

PROOF. Under the above conditions for A and F as well as condition (4.30a) for the perturbation $g(t,u)$, we are in the position to apply Theorem 1.4 to Eq. (4.30d). It yields that, for each initial value $u_0 \in L^\infty$ satisfying (4.30c), there exists a unique, global and classical solution u of (4.30d) such that $O(u) \subset [\tau_-, \tau_+]$ and that the section $K := \{u(t) : t \ge 1\}$ is contained and bounded in every $Y(\alpha,p)$. Particularly, K is bounded in $Y(\beta,q) = D(A^\beta \restriction L^q) \cap L^\infty$, which is compactly embedded in L^∞ by assumption. Therefore, K is precompact in L^∞. Moreover, since the inverse $(1+A)^{-1} : L^2 \to L^2$ is compact, the embedding $D(A^{3/4}) \hookrightarrow D(\sqrt{A}) = D(B)$ is compact. Thus K, as a bounded subset in $D(A^{3/4})$, is precompact in $D(\sqrt{A}) = D(B)$. Hence, the section K is precompact in $X = D(B) \cap L^\infty$.

Therefore, the convergence in (4.30f) follows from Theorem 1.2 (i) of this chapter, since this theorem is now applicable by virtue of the analyticity of the function F. □

Under the situation of Theorem 1.8, we are allowed to apply Proposition 1.5 to find that there exists a local ground state $\varphi \in \mathcal{S}_\tau$ such that

$$\mathcal{E}_F(\varphi) \le \mathcal{E}_F(u) \quad \forall u \in X_\tau. \tag{4.31a}$$

We are interested in the stability of this local ground state with respect to the corresponding autonomous gradient system

$$\begin{cases} \dot{u} = -\mathcal{E}_F'(u), \\ \mathcal{E}_F'(u) \equiv Au + f(x,u), \quad f \equiv \nabla_u F. \end{cases} \tag{4.31b}$$

Our tool for deriving the stability result is Theorem 4.10 of Chapter 3 corresponding to the following choices:

$$H := L^2(\Omega;\mathbb{R}^d), \quad X := D(\sqrt{A}) \cap L^\infty(\Omega;\mathbb{R}^d)$$

and the set $U = U_\tau \subset H$ is

$$U_\tau := \{w \in L^\infty(\Omega;\mathbb{R}^d) : \tau_- \le w \le \tau_+\}.$$

Note that $U_\tau \cap X = X_\tau$ is closed in X, and $\mathcal{E}_F : X_\tau \to \mathbb{R}$ attains its local minimum at $\varphi \in X_\tau$. Thus, in order to apply Theorem 4.10 of Chapter 3, we need to check the following properties (i)–(iii) for the solutions $u(\cdot;u_0)$ of the gradient system (4.31b) with initial data $u_0 \in U_\tau$:

(i) The solutions $u(\cdot;u_0)$ are such that

$$u \in C([0,\infty);L^2) \cap C^1((0,\infty);L^2) \cap C((0,\infty);X_\tau), \tag{4.31c}$$

and there exists a constant $c_1 > 0$ such that

$$\|u(1;u_0) - u(1;v_0)\|_{L^2} \le c_1 \cdot \|u_0 - v_0\|_{L^2} \quad \forall u_0, v_0 \in U_\tau. \tag{4.31d}$$

(ii) There exists a compact subset K of $X = D(\sqrt{A}) \cap L^\infty$ such that

$$\{u(t;u_0) : t \ge 1\} \subset K \quad \forall u_0 \in U_\tau. \tag{4.31e}$$

(iii) There exists a function $\phi \in \mathcal{G}$ such that

$$\phi(\mathcal{E}_F(w)) \le \|\mathcal{E}'_F(w)\|_{L^2} \quad \forall w \in K. \tag{4.31f}$$

The existence property in (i) follows immediately from Theorem 1.4 of this chapter. Moreover, as seen there, there exists a closed subset K of the Banach space $D(A^{3/2}) \cap Y(\beta, q)$, where the pair $(\beta, q) \in (0,1) \times [2, \infty)$ is such that the embedding $Y(\beta, q) \hookrightarrow L^\infty$ is compact, such that (4.31e) holds. Since both embeddings $Y(\beta, q) \hookrightarrow L^\infty$ and $D(A^{3/4}) \hookrightarrow D(\sqrt{A})$ are compact by our assumptions, the closed subset K of $D(A^{3/4}) \cap Y(\beta, q)$, when considered as a subset of $X = D(\sqrt{A}) \cap L^\infty$, is compact. As a consequence, by virtue of the uniform analyticity assumption on F, we find from an application of Theorem 1.1 (ii) of this chapter that there exists a function $\phi \in \mathcal{G}_2$ such that (4.31f) holds.

We still have to check the continuous dependence on the initial data given in (4.31d). To this end, we consider two solutions u and v of (4.31b) with initial data $u(0) = u_0 \in U_\tau$ and $v(0) = v_0 \in U_\tau$. By rewriting both solutions u and v as

$$u(t) = e^{-tA} u_0 + \int_0^t e^{-(t-s)A} f(\cdot, u(s))\, ds$$

and

$$v(t) = e^{-tA} v_0 + \int_0^t e^{-(t-s)A} f(\cdot, v(s))\, ds,$$

where $\{e^{-tA} : t \ge 0\}$ is the C_0 contractive semigroup on L^2 generated by the submarkovian operator A, we have, with $w(t) := u(t) - v(t)$, that

$$w(t) = e^{-tA} w(0) + \int_0^t e^{-(t-s)A} (f(\cdot, u(s)) - f(\cdot, v(s)))\, ds$$

for all $t \ge 0$. Since

$$\frac{d}{dr} f(\cdot, ru(s) + (1-r)v(s)) = F_2(\cdot, ru(s) + (1-r)v(s))(u(s) - v(s))$$

for all $r \in [0,1]$, where $F_2 := \mathrm{Hess}_u F$ (noting that $f = \nabla_u F$), we have that

$$f(\cdot, u(s)) - f(\cdot, v(s)) = \int_0^1 F_2(\cdot, ru(s) + (1-r)v(s)) w(s)\, dr.$$

The uniform boundedness of $u(s), v(s) \in [\tau_-, \tau_+]$ combined with the locally uniform boundedness of the map $L^\infty \ni w \mapsto F_2(\cdot, w) \in M_d(L^\infty)$ implies that there exists a constant $c_2 > 0$ such that

$$\|F_2(\cdot, ru(s) + (1-r)v(s))\|_{M_d(L^\infty)} \le c_2$$

for all $r \in [0,1]$, $s \geq 0$. Therefore, we have that

$$|f(\cdot, u(s)) - f(\cdot, v(s))| \leq c_2 \cdot |w(s)|.$$

It follows that

$$\|f(\cdot, u(s)) - f(\cdot, v(s))\|_{L^2} \leq c_2 \cdot \|w(s)\|_{L^2}$$

as well as

$$\|e^{-(t-s)A}(f(\cdot, u(s)) - f(\cdot, v(s)))\|_{L^2} \leq c_2 \cdot \|w(s)\|_{L^2}$$

for all $s \in [0,t], t \geq 0$, since the semigroup $\{e^{-tA} : t \geq 0\}$ is contractive on L^2. We have further that $\|e^{-tA}w(0)\|_{L^2} \leq \|w(0)\|_{L^2}$ for all $t \geq 0$. Hence, there holds

$$\|w(t)\|_{L^2} \leq \|w(0)\|_{L^2} + c_2 \int_0^t \|w(s)\|_{L^2}\, ds \quad (t \geq 0).$$

Now an application of the classical Gronwall inequality yields finally that

$$\|w(t)\|_{L^2} \leq (1 + (e^{c_2 t} - 1)/c_2) \cdot \|w(0)\|_{L^2}$$

for all $t \geq 0$. Hence, (4.31d) holds true with the constant $c_1 := 1 + e^{c_2}/c_2$.

The above shows that all conditions in Theorem 4.10 of Chapter 3 are satisfied with $H := L^2$, $X := D(\sqrt{A}) \cap L^\infty$ and $U = U_\tau$. Therefore, an application of this theorem yields the following stability result.

THEOREM 1.9. **(Stability of ground state)**

Let the above conditions given in the boxes for the operator A and the nonlinearity F be satisfied. Assume furthermore that the function $F : \Omega \times \mathbb{R}^d \to \mathbb{R}$ is uniformly analytic with respect to the second argument.

Let $\varphi \in S_\tau$ be a local ground state, i.e.,

$$\mathcal{E}_F(\varphi) \leq \mathcal{E}_F(u) \quad \forall u \in X_\tau.$$

Then φ is Lyapunov stable in the sense that for any given $\varepsilon > 0$, there exists a constant $\delta \equiv \delta(\varepsilon) > 0$ with the following property: For any initial value $u_0 \in L^\infty$, $\tau_- \leq u_0 \leq \tau_+$, with

$$\|u_0 - \varphi\|_{L^2} < \delta,$$

there holds

$$\sup_{t \geq 1} \|u(t) - \varphi\|_X < \varepsilon,$$

where u is the unique, global and classical solution of the gradient system (4.31b) with $u(0) = u_0$.

Moreover, if φ is an isolated critical point of $\mathcal{E}_F$ in X_τ, then φ is uniformly asymptotically stable in the sense that there exists a constant $\delta_0 > 0$ such that

$$\|u(t) - \varphi\|_X \to 0 \quad \text{as} \quad t \to \infty$$

uniformly for all solutions u of (4.31b) with initial data $u_0 \in L^\infty$ satisfying $\tau_- \leq u_0 \leq \tau_+$ and $\|u_0 - \varphi\|_{L^2} < \delta_0$.

1.6. Remarks. (a) In the proofs of Lemma 1.3 as well as Theorem 1.4 of this chapter, we did not use the fact that f is the gradient of the function F, except for the energy equality (4.22d). What we have used for the proof of the existence of global solutions satisfying (4.22b,c,e) is essentially the growth conditions (4.18a,b,c) for the map f. Hence, these results (4.22b,c,e) remain valid for any map f satisfying conditions (4.18a,b,c).

(b) If each L_j $(j = 1, 2, ..., d)$ is a symmetric strongly elliptic operator on a bounded domain Ω, then $A := -\text{diag}(L_1, ..., L_d)$ has the submarkovian property and the compactness described in ♠. Thus the results obtained above apply to these operators.

(c) In many application situations, only **positive solutions** will be taken into account. Note that the growth condition (4.18c) for the map f is needed only for Lemma 1.3. Therefore, if we are working in positive solutions, then it is sufficient to have a growth condition of the form (4.18c) for all positive u. In this case, the assertions in (4.22b,c,e) of Theorem 1.4 remain valid for positive solutions with initial data u_0 satisfying (4.22a).

(d) As seen in Proposition 1.5, the growth conditions (4.18a,b,c) ensure the existence of solutions to the equation $\mathcal{E}'_F(u) = Au + f(x, u) = 0$, provided that A is a diagonal, positive semidefinite symmetric submarkovian generator such that the embedding $D(\sqrt{A}) \cap L^\infty \hookrightarrow L^2$ is compact.

To explain the situation discussed in (c), we give an example.

EXAMPLE 1.10. **(Population models for two species).** The simplest reaction-diffusion models (see Smoller [**110**]) have the form

$$(RD) \qquad \begin{cases} \frac{\partial u}{\partial t} = d_1 \Delta u - u(u - a_1 - a_2 v), \\ \frac{\partial v}{\partial t} = d_2 \Delta v - v(v - b_1 - b_2 u), \\ \frac{\partial u}{\partial \mathbf{n}} = 0 = \frac{\partial v}{\partial \mathbf{n}} \quad \text{on} \quad \partial\Omega. \end{cases}$$

Here $u \equiv u(x, t)$ and $v \equiv v(x, t)$ are the population densities. d_1, d_2 are the diffusion parameters, which are positive constants. The "no flux" boundary condition reflects an ecological situation such that the species is surrounded by the bounded domain $\Omega \subset \mathbb{R}^N$, from which they cannot escape.

There are also four constants a_1, a_2, b_1, b_2. The signs of a_2 and b_2 describe the different model types (Smoller [**110**, pp. 218–221]):

$$\begin{cases} \text{Predator-Prey model:} \quad \text{where} \quad a_2 < 0, b_2 > 0; \\ \text{Competing-Species model:} \quad \text{where} \quad a_2 < 0, b_2 < 0; \\ \text{Symbiosis model:} \quad \text{where} \quad a_2 > 0, b_2 > 0. \end{cases}$$

Observe that (RD) can be rewritten as

$$(RD') \qquad \dot{w} = -Aw - f(w), \quad w := (u, v)^T,$$

where $A := \text{diag}(-d_1 \Delta_N, -d_2 \Delta_N)$ (here Δ_N is the Laplacian assigned to the Neumann (no flux) boundary condition), and $f(w) \equiv (f_1(w), f_2(w))$ with

$$f_1(w) := u(u - a_1 - a_2 v), \quad f_2(w) := v(v - b_1 - b_2 u).$$

Clearly, A is a submarkovian generator and, by the Sobolev Embedding Theorem, the embeddings

$$D(\sqrt{A} \restriction L^q) \subset W^{1,q}(\Omega;\mathbb{R}^2) \hookrightarrow C(\bar{\Omega};\mathbb{R}^2)$$

are compact for all $q > N$.

Since (RD') is an unperturbed system and we are interested only in positive solutions, we need to consider the growth condition (4.18c) corresponding to the choices $M = 0$ and $\tau_- = 0$. Since $f_1(0,s) = 0 = f_2(s,0)$ for all $s \in \mathbb{R}$, condition (4.18c) has been satisfied for the choices $M = 0$ and $\tau_- = 0$. Hence, we are left to find conditions on the parameters a_j's and b_j' that ensure the existence of two positive constants τ_1, τ_2 such that condition (4.18c) is satisfied with $M := 0, \tau_- := 0$ and $\tau_+ := (\tau_1, \tau_2)$, i.e.,

$$(*) \qquad \begin{cases} f_1(\tau_1, v) \geq 0 & \forall 0 \leq v \leq \tau_2; \\ f_2(u, \tau_2) \geq 0 & \forall 0 \leq u \leq \tau_1. \end{cases}$$

Equivalently,

$$(**) \qquad a_2 v \leq \tau_1 - a_1 \quad \forall 0 \leq v \leq \tau_2; \qquad b_2 u \leq \tau_2 - b_1 \quad \forall 0 \leq u \leq \tau_1.$$

It is now easy to see that both inequalities in $(**)$ will be satisfied for sufficiently large $\tau_{1,2}$, provided the parameters a_2, b_2 are such that

$$(***) \qquad \text{either} \quad a_2 \leq 0, b_2 \leq 0 \quad \text{or} \quad a_2 b_2 < 1.$$

Note that $(***)$ is automatically satisfied for both the predator-prey model and the competing-species model. Hence, only for the symbiosis model does it become a real condition.

In conclusion, we have shown that condition $(***)$ ensures that the RD-system (4.31) has invariant domains of the form $\{(u,v) : 0 \leq u \leq \tau_1, 0 \leq v \leq \tau_2\}$ for all sufficiently large $\tau_{1,2}$ satisfying $(**)$. Moreover, as pointed out in Remark 1.6 (c), for these positive solutions $w = (u,v)$ of (RD) with positive and bounded initial data we have the precompactness of the section $\{w(t) : t \geq 1\}$ in the subspace

$$X = D(\sqrt{A}) \cap L^\infty \subset W^{1,2}(\Omega;\mathbb{R}^2) \cap L^\infty.$$

In other words, the reaction-diffusion system (RD) has a compact attractor in $W^{1,2}(\Omega;\mathbb{R}^2) \cap L^\infty$ that attracts every bounded and positive solution of (RD).

More general results in this topic can be found in [**23, 110, 125**]. It should be kept in mind that it is generally unknown whether every bounded and positive solution of (RD) converges to some equilibrium.

2. The perfect case

As seen in §1, the "natural" choice of the definition domain of the functional $\mathcal{E}_F$ given by (4.5b) is the Banach space $X = V \cap L^\infty$ with $V = D(B) = D(\sqrt{A})$. In general, however, this choice is far from optimal. In fact, the best possible choice of the definition domain for $\mathcal{E}_F$ is $V = D(B)$, the definition domain of the linear part. Certainly, such a perfect case occurs only under (sometimes very) strict growth restrictions on the analytic function F, and such a growth restriction will usually force the function F to be a polynomial. However, this does not completely destroy our interest in developing an applicable theory for such a perfect case. In fact, many systems treated in the applied sciences (cf. Cross-Hohenberg [**26**]), as approximations of complicated systems, have polynomial nonlinearity F and fit such a perfect case.

Below we present an abstract theory, and then we will give its concrete applications in §3.

We consider a Hilbert space $\mathcal{H}$ and assume that

$$B : D(B) \subset \mathcal{H} \to \mathcal{H} \tag{4.32a}$$

is a self-adjoint and positive definite linear operator such that its inverse

$$B^{-1} : \mathcal{H} \to \mathcal{H} \quad \text{is compact.} \tag{4.32b}$$

The domain

$$V := (D(B), \|\cdot\|_B) \tag{4.32c}$$

becomes a Hilbert space under the inner product

$$\langle u, v\rangle_V := \langle Bu, Bv\rangle_{\mathcal{H}} \quad (u, v \in D(B)). \tag{4.32d}$$

We identify the dual space V' with the extrapolated Hilbert space $\mathcal{H}_{-1}$ associated with B. It follows that

$$\|f\|_{V'} = \|B^{-1}f\|_{\mathcal{H}} \quad \forall f \in \mathcal{H}. \tag{4.32e}$$

Consequently, both embeddings

$$V \hookrightarrow \mathcal{H} \hookrightarrow V'$$

are compact.

Let $\mathcal{F} : V \to \mathbb{R}$ be a functional such that

$$\mathcal{K} = \mathcal{F}' : V \to \mathcal{H} \tag{4.33a}$$

is an **analytic** gradient mapping that maps bounded subsets of V into bounded subsets of $\mathcal{H}$. It follows from the compactness of the embedding $\mathcal{H} \hookrightarrow V'$ that

$$\mathcal{K} : V \to V' \quad \text{is analytic and compact.} \tag{4.33b}$$

Consequently, a result of M. A. Krasnoselski [**75**, Theorem 8.1, p. 57] implies that, for each $u \in V$, the linear operator

$$\mathcal{K}'(u) \in L(V, V') \quad \text{is compact.} \tag{4.33c}$$

Consider the functional

$$\mathcal{E}_{\mathcal{F}}(u) := \frac{1}{2}\|Bu\|_{\mathcal{H}}^2 + \mathcal{F}(u) \quad (u \in V). \tag{4.34a}$$

Then $\mathcal{E}_{\mathcal{F}}$ is an analytic functional with

$$\mathcal{E}_{\mathcal{F}}'(u) = B^2u + \mathcal{K}(u) \in V', \quad \mathcal{E}_{\mathcal{F}}''(u) = B^2 + \mathcal{K}'(u) \in L(V, V') \tag{4.34b}$$

for all $u \in V$. Let

$$\mathcal{S} := \{u \in V : B^2 u + \mathcal{K}(u) = 0 \quad \text{in} \quad V'\} \tag{4.34c}$$

be the set of critical points of $\mathcal{E}_{\mathcal{F}}$.

Let $\varphi \in \mathcal{S}$. Since

$$B^2 : V \to V'$$

is an isomorphism from V onto V' and $\mathcal{K}'(\varphi) \in L(V, V')$ is compact, it follows from the classical Riesz-Schauder theory that the linear operator $\mathcal{E}''_{\mathcal{F}}(\varphi) = B^2 + \mathcal{K}'(\varphi) \in L(V, V')$ is a Fredholm operator of index zero. Therefore, we are in the position to apply Theorem 4.5 of Chapter 2 (of Huang and Takáč [**70**]). It yields the following gradient inequality:

$$c \cdot |\mathcal{E}_{\mathcal{F}}(u) - \mathcal{E}_{\mathcal{F}}(\varphi)|^{\theta} \leq \|\mathcal{E}'_F(u)\|_{V'} \quad \forall u \in X, \|u - \varphi\|_V < \sigma, \tag{4.35}$$

where $c, \sigma, \theta \in [1/2, 1)$ are constants.

To improve this result, we need the following auxiliary result.

LEMMA 2.1. *Let $\{u_n\} \subset V$ be a bounded sequence in V such that*

$$\|B^2 u_n + \mathcal{K}(u_n)\|_{V'} \to 0 \quad \textit{as} \quad n \to \infty. \tag{4.36}$$

Then the set $\{u_n : n \in \mathbb{N}\}$ is precompact in V.

As a consequence, the set $\mathcal{S}$ is locally compact in the sense that for each $R > 0$, the subset

$$\mathcal{S}_R := \{u \in \mathcal{S} : \|u\|_V \leq R\}$$

is compact in V.

PROOF. It suffices to show that the given sequence $\{u_n\} \subset V$ contains a convergent subsequence in V. This can be done as follows.

The compactness of the map $\mathcal{K} : V \to V'$ (see (4.33b)) implies that $\{\mathcal{K}(u_n)\}$ contains a subsequence that is convergent in V'. Thus, we may assume by passing to a subsequence if necessary that $\|\mathcal{K}(u_n) - g\|_{V'} \to 0$ with some $g \in V'$. Hence, by taking the limit in (4.36) we find that

$$\|B^2 u_n + g\|_{V'} \to 0$$

and thus

$$\|u_n + B^{-1} g\|_V = \|B^2 u_n + g\|_{V'} \to 0$$

as $n \to \infty$. This completes the proof. □

PROPOSITION 2.2. *For each $R > 0$ there exists some function $\phi \in \mathcal{G}_2$ such that*

$$\phi(\mathcal{E}_{\mathcal{F}}(u)) \leq \|\mathcal{E}'_{\mathcal{F}}(u)\|_{V'} \quad \forall u \in V, \|u\|_V \leq R. \tag{4.37}$$

PROOF. By Lemma 2.1, the set $K := \mathcal{S}_R$ is compact. As seen above, $\mathcal{E}'_{\mathcal{F}}$ satisfies a gradient inequality of LS-type near every $u \in K$. Therefore, for each $u \in K$ we can find a neighborhood $K(u) \subset V$ of u and two constants $c(u) \in (0, 1/2]$, $\theta(u) \in [1/2, 1)$ such that

$$c(u) \cdot |\mathcal{E}_{\mathcal{F}}(u) - \mathcal{E}_{\mathcal{F}}(v)|^{\theta(u)} \leq \|\mathcal{E}'_{\mathcal{F}}(v)\|_{V'} \quad \forall v \in K(u).$$

Now $\{K(u) : u \in K\}$ forms an open cover of K. Hence, the compactness of K yields a finite cover $\{K(u_i)\}_{i=1}^{i=N}$. For each i, $1 \leq i \leq N$, set

$$\phi_i(x) := c(u_i) \cdot \min\{|x - \mathcal{E}_{\mathcal{F}}(u_i)|^{\theta(u_i)}, |x - \mathcal{E}_{\mathcal{F}}(u_i)|^{1/2}\} \quad (x \in \mathbb{R}).$$

Each ϕ_i belongs to the class $\mathcal{G}_2$. It follows that

$$\phi_0(x) := \min\{\phi_i(x) : 1 \leq i \leq N\} \quad (x \in \mathbb{R})$$

is of class $\mathcal{G}_2$, and if $u \in K(u_i)$ for some $1 \leq i \leq N$, then we have

$$\phi_0(\mathcal{E}_{\mathcal{F}}(u)) \leq \phi_i(\mathcal{E}_{\mathcal{F}}(u)) \leq c(u_i) \cdot |\mathcal{E}_{\mathcal{F}}(u) - \mathcal{E}_{\mathcal{F}}(u_i)|^{\theta(u_i)} \leq \|\mathcal{E}'_{\mathcal{F}}(u)\|_{V'}.$$

Consequently, in the neighborhood

$$\tilde{K} := \bigcup_{i=1}^{N} K(u_i) \supset K \tag{4.38a}$$

we have

$$\phi_0(\mathcal{E}_{\mathcal{F}}(u)) \leq \|\mathcal{E}'_{\mathcal{F}}(u)\|_{V'} \quad \forall u \in \tilde{K}. \tag{4.38b}$$

We want to show that there exists some constant $c > 0$ such that the expected gradient inequality (4.37) holds true under the choice of the function $\phi := c \cdot \phi_0 \in \mathcal{G}_2$.

We argue by contradiction. If this were false, then for each $n \in \mathbb{N}$ we can find some $u_u \in V$ such that

$$\|u_n\|_V \leq R \quad \text{but} \quad \frac{1}{n}\phi_0(\mathcal{E}_{\mathcal{F}}(u_n)) > \|\mathcal{E}'_{\mathcal{F}}(u_n)\|_{V'}. \tag{4.39a}$$

Since $\phi_0 \in \mathcal{G}_2$, we have

$$\phi_0(\mathcal{E}_{\mathcal{F}}(u_n)) \leq 2\phi_0(0) + |\mathcal{E}_{\mathcal{F}}(u_n)|^{1/2}. \tag{4.39b}$$

The boundedness of the sequence $\{u_n\} \subset V$ implies the boundedness of the sequence $\{\mathcal{E}_{\mathcal{F}}(u_n)\}$. Hence, by combining (4.39a) with (4.39b), we find that

$$\|B^2 u_n + \mathcal{K}(u_n)\|_{V'} = \|\mathcal{E}'_{\mathcal{F}}(u_n)\|_{V'} \to 0.$$

Therefore, the bounded sequence $\{u_n\} \subset V$ satisfies condition (4.36) of Lemma 2.1 and thus an application of Lemma 2.1 yields that $\{u_n\} \subset V$ contains some subsequence that converges to some $\varphi \in \mathcal{S}_R = K$. Hence, by passing to a subsequence if necessary, we may assume that $\{u_n\}$ is contained in the neighborhood $\tilde{K}$ of K and thus, we are allowed to apply (4.38b) to each u_n. It yields that $\phi_0(\mathcal{E}_{\mathcal{F}}(u_n)) \leq \|\mathcal{E}'_{\mathcal{F}}(u_n)\|_{V'}$ and thus

$$\frac{1}{n}\phi_0(\mathcal{E}_{\mathcal{F}}(u_n)) > \|\mathcal{E}'_{\mathcal{F}}(u)\|_{V'} \geq \phi_0(\mathcal{E}_{\mathcal{F}}(u_n))$$

for all $n \geq 1$. Clearly, this is impossible. □

To complete the setup, we impose the final conditions concerning the coercivity of the functional $\mathcal{E}_{\mathcal{F}}$ and the decay of the perturbation g:

(i) For each $R > 0$, the set

$$\{u \in V : \mathcal{E}_{\mathcal{F}}(u) \leq R\} \quad \text{is bounded in} \quad V. \tag{4.40a}$$

(ii) The map $g : \mathbb{R}_+ \times V \to \mathcal{H}$ is locally Lipschitz in the second argument and locally Hölder continuous in time. Moreover, it satisfies the following smallness condition:

$$\text{(4.40b)} \qquad \begin{cases} \|g(t,u)\|_{\mathcal{H}} \le g(t) \quad \forall (t,u) \in \mathbb{R}_+ \times V, \\ g \in L^\infty \quad \text{and} \quad \displaystyle\int_0^\infty \rho(t) g(t)^2\,dt + \int_0^\infty \frac{1}{\rho(t)}\,dt < \infty \\ \text{with some increasing positive function} \quad \rho. \end{cases}$$

Under the above circumstances we have the following gradient-like system:

$$\text{(4.41a)} \qquad \begin{cases} \dot{u}(t) = -\mathcal{E}'_{\mathcal{F}}(u(t)) + g(t,u(t)), \\ \mathcal{E}'_{\mathcal{F}}(u) \equiv B^2 u + \mathcal{K}(u), \quad \mathcal{K} \equiv \mathcal{F}'. \end{cases}$$

Our convergence result reads as follows.

THEOREM 2.3. *Assume conditions (4.32)–(4.33) as well as (4.40). Then, for any initial value $u_0 \in V$, the Cauchy problem for (4.41a) has a unique global classical solution*

$$\text{(4.41b)} \qquad u \in C([0,\infty);V) \cap C((0,\infty);D(B^2)) \cap C^1((0,\infty);\mathcal{H})$$

such that

$$\text{(4.41c)} \qquad \lim_{t\to\infty} \|u(t) - \varphi\|_V = 0$$

with some equilibrium $\varphi \in \mathcal{S}$.

PROOF. The operator $A := B^2$ is also self-adjoint and positive definite in the Hilbert space $\mathcal{H}$. Moreover, there holds $\sqrt{A} = B$ and thus $D(\sqrt{A}) = D(B) = V$. In condition (4.33a) we have assumed that the map $\mathcal{K} = \mathcal{F}' : V \to \mathcal{H}$ is analytic and thus it is particularly locally Lipschitz continuous. Furthermore, we have assumed that the perturbation $g(t,v)$ is locally Hölder continuous in time and locally Lipschitz continuous for $v \in V$. Therefore, we are in the position to apply Theorem 3.3.3 of Henry [**63**, p. 54] (with the choice $\alpha = \frac{1}{2}$). It yields that for every $u_0 \in V$, the Cauchy problem (4.41a) admits a unique classical solution

$$\text{(4.42a)} \qquad u \in C([0,T);V) \cap C((0,T);D(B^2)) \cap C^1((0,T);\mathcal{H})$$

with the maximal existence interval $[0,T)$, and

$$\text{(4.42b)} \qquad \limsup_{t\to T} \|u(t)\|_V = \infty \quad \text{if} \quad T < \infty.$$

Moreover, there holds the energy equality

$$\text{(4.42c)} \qquad \frac{d}{dt}\mathcal{E}_{\mathcal{F}}(u(t)) = \langle \mathcal{E}'_{\mathcal{F}}(u(t)), \dot{u}(t)\rangle_{\mathcal{H}} \quad \forall t \in (0,T).$$

To show $T = \infty$ and the convergence in (4.41c), we use the equation $\mathcal{E}'_{\mathcal{F}}(u) = g(t,u) - \dot{u}$ to obtain that

$$\langle \mathcal{E}'_{\mathcal{F}}(u(t)), \dot{u}(t)\rangle_{\mathcal{H}} = \langle g(t,u(t)) - \dot{u}(t), \dot{u}(t)\rangle_{\mathcal{H}} = -f(s)^2 + g_1(t)^2$$

with

$$(4.43a) \qquad \begin{cases} f(t) := \|\dot{u}(t) + g(t,u(t))/2\|_{\mathcal{H}}, \\ g_1(t) := \frac{1}{2}\|g(t,u(t))\|_{\mathcal{H}} \end{cases} \qquad \forall t \in (0,T).$$

Therefore, by integrating (4.42c), we find that

$$(4.43b) \qquad \mathcal{E}_{\mathcal{F}}(u_0) - \mathcal{E}_{\mathcal{F}}(u(t)) = \int_0^t f(s)^2\,ds - \int_0^t g_1(s)^2\,ds.$$

Note that

$$0 \le g_1(t) \le g(t)$$

by (4.40b). We find from (4.43b) that

$$\mathcal{E}_{\mathcal{F}}(u(t)) \le \mathcal{E}_{\mathcal{F}}(u_0) + \|g\|_{L^2} < \infty$$

by noting that $g \in L^2(\mathbb{R}_+)$ (see (4.40b) again). Hence, the orbit $O(u) = \{u(t) : t \in [0,T)\}$ is contained in the set $\{v \in V : \mathcal{E}_{\mathcal{F}}(v) \le \mathcal{E}_{\mathcal{F}}(u_0) + \|g\|_{L^2}\}$ that is bounded by condition (4.40a). By virtue of (4.42b), this implies that $T = \infty$ and hence the orbit $O(u) = \{u(t) : t \ge 0\}$ is a bounded subset of V. Now the sectorial operator $A := B^2$ has compact resolvent and the map $\mathcal{K}(t,u) := -\mathcal{K}(u) + g(t,u)$ has the property that it maps the sets $\mathbb{R}_+ \times U$ with bounded subsets U of V into bounded subsets of $\mathcal{H}$, by noting that $g \in L^\infty(\mathbb{R}_+)$. Hence, Theorem 3.3.6 of Henry [**63**, p. 57] applies and yields that the orbit $O(u) = \{u(t) : t \ge 0\}$ is precompact in $D(\sqrt{A}) = V$.

As seen before, the analytic gradient map $\mathcal{E}'_{\mathcal{F}} : V \to V'$ satisfies a gradient inequality of LS-type near every point. Hence, we are allowed to apply Theorem 4.2 (ii) of Chapter 3 corresponding to the choice of the pair $(H,X) = (\mathcal{H},V)$ to derive the desired convergence in (4.41c). This completes the proof. □

By setting $g \equiv 0$ in (4.41a), we have the following autonomous gradient system

$$(4.44a) \qquad \dot{u} = -\mathcal{E}'_{\mathcal{F}}(u).$$

The following stability result says that the functional $\mathcal{E}_{\mathcal{F}}$ has an absolute ground state which, as an equilibrium of the gradient system (4.44a), is stable.

THEOREM 2.4. **(Existence and stability of ground states)**
Under the circumstances of Theorem 2.3 there exists some $\varphi \in V = D(B)$ such that

$$(4.44b) \qquad \mathcal{E}_{\mathcal{F}}(\varphi) \le \mathcal{E}_{\mathcal{F}}(u) \quad \forall u \in V.$$

Moreover, this ground state φ is Lyapunov stable in the sense that for any given $R > \mathcal{E}_{\mathcal{F}}(\varphi)$ and $\varepsilon > 0$, there exists a constant $\delta \equiv \delta(R,\varepsilon) > 0$ with the following property: For any initial value $u_0 \in V$ with

$$\mathcal{E}_{\mathcal{F}}(u_0) \le R, \quad \|u_0 - \varphi\|_{\mathcal{H}} < \delta$$

there holds

$$\sup_{t\geq 1}\|u(t)-\varphi\|_V<\varepsilon,$$

where u is the unique, global and classical solution of the gradient system (4.44a) with the initial value $u(0)=u_0$.

Moreover, if φ is an isolated critical point of $\mathcal{E}_{\mathcal{F}}$, then φ is uniformly asymptotically stable in the sense that for each $R>\mathcal{E}_{\mathcal{F}}(\varphi)$, there exists some constant $\delta_0>0$ such that

$$\|u(t)-\varphi\|_V\to 0 \quad as \quad t\to\infty$$

uniformly for all solutions u of (4.44a) with initial data $u_0\in V$ satisfying $\mathcal{E}_{\mathcal{F}}(u_0)\leq R$ and $\|u_0-\varphi\|_{\mathcal{H}}<\delta_0$.

PROOF. We show the existence of the ground state first. To this end, we choose $R>0$ to be sufficiently large so that the set

$$\Sigma:=\{u\in V:\mathcal{E}_{\mathcal{F}}(u)\leq R\} \tag{4.45}$$

is non-empty. Σ is bounded by our assumption (see (4.40a)). This implies that

$$\alpha:=\inf\{\mathcal{E}_{\mathcal{F}}(u):u\in X\}=\inf\{\mathcal{E}_{\mathcal{F}}(u):u\in\Sigma\}>-\infty. \tag{4.46}$$

We choose a sequence $\{u_n\}\subset\Sigma$ such that

$$\mathcal{E}_{\mathcal{F}}(u_n)\to\alpha \quad \text{as} \quad n\to\infty. \tag{4.47}$$

We apply Theorem 2.3 under the special choice $g\equiv 0$. It follows that each initial value u_n generates a classical global solution $u_n(t)$ such that

$$\alpha\leq\mathcal{E}_{\mathcal{F}}(u_n(t))\leq\mathcal{E}_{\mathcal{F}}(u_n) \quad \forall t>0,$$

and there exists a $\varphi_n\in\mathcal{S}_R$ such that

$$\|u_n(t)-\varphi_n\|_V\to 0 \quad \text{as} \quad n\to\infty.$$

Therefore, the sequence $\{\varphi_n\}\subset\mathcal{S}_R$ is such that

$$\mathcal{E}_{\mathcal{F}}(\varphi_n)\to\alpha \quad \text{as} \quad n\to\infty. \tag{4.48}$$

Since the set $\mathcal{S}_R$ is compact in V by Lemma 2.1, $\{\varphi_n\}$ contains a cluster point, say φ, for which we have from (4.48) that $\alpha=\mathcal{E}_{\mathcal{F}}(\varphi)$. Hence, φ is a ground state for $\mathcal{E}_{\mathcal{F}}$.

We derive the above-mentioned stability of the ground state φ by virtue of Theorem 4.10 of Chapter 3 under the following choices of the spaces:

$$H:=\mathcal{H}, \quad X:=V,$$

as well as $U=U_R$, where the set U_R for each $R>\mathcal{E}_{\mathcal{F}}(\varphi)$ is given by $U_R:=\{v\in V:\mathcal{E}_{\mathcal{F}}(v)\leq R\}$. Remember that in order to apply Theorem 4.10 of Chapter 3, we must check that the corresponding existence, continuous dependence on initial data, and compactness of the solutions of (4.44a) (cf. (3.48a,b,c)) are satisfied. Because the checking of these three properties is very similar to that for Theorem 1.8 of this chapter, we omit the details. Note that each set U_R is invariant for the gradient system (4.44a) and the previous Proposition 2.2 implies that the gradient map $\mathcal{E}'_{\mathcal{F}} : V \to V'$ satisfies a gradient inequality in every U_R. Thus, the stability results mentioned in the theorem follow from an application of Theorem 4.10 of Chapter 3. □

3. Results around the Laplacian operators

In the following, $\Omega \subset \mathbb{R}^N$ is a bounded domain with a Lipschitz boundary $\partial\Omega$, and d is a positive integer. We will present the results for both the scalar case ($d = 1$) and the vectorial case ($d > 1$) in a unified way.

Let $F : \Omega \times \mathbb{R}^d \to \mathbb{R}$ be uniformly analytic with respect to the second argument, and put

$$\mathcal{E}_F(u) := \sum_{j=1}^{d} \frac{\alpha_j}{2} \int_\Omega |\nabla u_j|^2 \, dx + \int_\Omega F(x, u) \, dx, \tag{4.49a}$$

where $\alpha_1, ..., \alpha_d$ are positive constants. The definition domain of $\mathcal{E}_F$ is contained in the definition domain of the linear part; i.e., in general, we have

$$\mathrm{dom}(\mathcal{E}_F) \subset W^{1,2}(\Omega; \mathbb{R}^d). \tag{4.49b}$$

The Fréchet derivative is given formally by

$$\mathcal{E}'_F(u) = -\alpha \Delta u + f(x, u), \tag{4.49c}$$

where

$$\alpha := \mathrm{diag}(\alpha_1, ..., \alpha_d), \quad f \equiv \nabla_u F. \tag{4.49d}$$

We take

$$\begin{cases} \mathcal{H} := L^2(\Omega; \mathbb{R}^d), & A := \mathrm{diag}(A_1, ..., A_d) \quad \text{with} \\ A_j := -\alpha_j \Delta \quad (j = 1, ..., d), \end{cases} \tag{4.50a}$$

and the Laplacian is assigned with one of the following classical boundary conditions:

(B1) (Dirichlet boundary condition)

$$D(\Delta) = W^{2,2}(\Omega) \cap W_0^{1,2}(\Omega).$$

(B2) (Neumann boundary condition)

$$D(\Delta) = \{u \in W^{2,2}(\Omega) : \frac{\partial u}{\partial \mathbf{n}} = 0 \quad \text{on} \quad \partial\Omega\}.$$

(B3) (Robin boundary condition)

$$D(\Delta) = \{u \in W^{2,2}(\Omega) : \frac{\partial u}{\partial \mathbf{n}} + \beta(x)u = 0 \quad \text{on} \quad \partial\Omega\},$$

where $\beta(x)$ is a positive function on $\partial\Omega$.

(B4) (Periodic boundary condition)

$$\Omega = [0, l_1] \times \cdots \times [0, l_N] \subset \mathbb{R}^N \quad \text{and}$$
$$D(\Delta) = \{u \in W^{2,2}(\Omega) : u \quad \text{is} \quad \Omega\text{-periodic}\}.$$

Under the above boundary conditions, each operator

$$A_j = -\alpha_j \Delta : D(\Delta) \subset L^2(\Omega; \mathbb{R}) \to L^2(\Omega; \mathbb{R})$$

is self-adjoint and positive semidefinite, and the square root

$$B := A^{1/2} = (A_1^{1/2}, ..., A_d^{1/2})$$

has the definition domain

$$V := D(B) \subset W^{1,2}(\Omega; \mathbb{R}^d) \tag{4.50b}$$

whose dual is contained in $W^{-1,2}(\Omega; \mathbb{R}^d)$:

$$V' \subset W^{-1,2}(\Omega; \mathbb{R}^d). \tag{4.50c}$$

Moreover, there holds

$$D(B \restriction L^p) \subset W^{1,p}(\Omega; \mathbb{R}^d) \quad (p \in [2, \infty)). \tag{4.50d}$$

Using the operator B, we rewrite the above functional $\mathcal{E}_F$ as

$$\mathcal{E}_F(u) = \frac{1}{2}\|Bu\|_{L^2}^2 + \int_\Omega F(x, u)\, dx. \tag{4.51}$$

$\mathcal{E}_F$ is in the form with which we have dealt in §1 and §2, and thus the abstract results obtained there will be applicable whenever the other related conditions are satisfied.

As seen in §1 and §2, two kinds of results are available: Those concerning the generic case where the Banach space $X \equiv V \cap L^\infty$ is chosen as the definition domain of $\mathcal{E}_F$ and those concerning the perfect case where the definition domain of $\mathcal{E}_F$ is just V. We will treat these two cases separately.

3.1. The generic case. We need to show that the diagonal operator $A = -\mathrm{diag}(\alpha_1\Delta, ..., \alpha_d\Delta)$ satisfies the conditions given in the box before Theorem 1.8, i.e., A is a positive semidefinite symmetric submarkovian generator such that the inverse $(1 + A)^{-1} : L^2 \to L^2$ is compact, and there exist some $\beta \in (0, 1)$ and $q \in [2, \infty)$ such that the embedding $D(A^\beta \restriction L^q) \cap L^\infty \hookrightarrow L^\infty$ is compact (we will show that this is true with $\beta := 1/2$ and any $q > N$).

The markovian property of the Laplacian is well known, cf., e.g., Pazy [**94**, Theorem 3.6, p. 215] or Henry [**63**, p. 32]. A direct verification of this result, i.e., the check of the equivalent Beurling-Deny characterization condition (4.17a′) goes as follows. Note that

$$\operatorname{div}((v-1)^+\nabla v) = \begin{cases} (v-1)^+\Delta v + |\nabla v|^2 & \text{if} \quad v > 1; \\ (v-1)^+\Delta v & \text{if} \quad v \le 1 \end{cases}$$

(cf. Gilbarg-Trudinger [**45**, Lemma 7.6, p. 32] or Ziemer [**126**, Cor. 2.18, p. 47]). It follows from the divergence theorem (see Gilbarg-Trudinger [**45**, p. 13]) that

$$\int_\Omega (\Delta v)(v-1)^+\, dx = \int_{\partial\Omega} (v-1)^+ \frac{\partial v}{\partial \mathbf{n}}\, dS - \int_{\{v>1\}} |\nabla v|^2\, dx \le 0,$$

since the integrand in the first integral is non-negative under each of the given boundary conditions. This implies for all $u = (u_1, ..., u_d) \in D(A)$ that

$$\begin{aligned}\langle Au, (u-1)^+\rangle_{L^2} &= \sum_{j=1}^{d} \langle A_j u_j, (u_j - 1)^+\rangle_{L^2} \\ &= \sum_{j=1}^{d} \alpha_j \int_\Omega (-\Delta u_j)(u_j - 1)^+ \, dx \geq 0.\end{aligned}$$

Hence, A is a positive semidefinite symmetric submarkovian generator.

The compactness of the inverse $(1+A)^{-1} : L^2 \to L^2$ is clear. Note that by (4.50d) we have that

$$D(B \upharpoonright L^q) \cap L^\infty \subset W^{1,q}(\Omega; \mathbb{R}^d)$$

for each $q \in [2, \infty)$. By the Sobolev Embedding Theorem (see Adams [**1**, Theorem 5.4]), the embeddings

$$W^{1,q}(\Omega; \mathbb{R}^d) \hookrightarrow C(\bar{\Omega}; \mathbb{R}^d)$$

are compact for all $q > N$. In particular, the imbedding

$$D(B \upharpoonright L^{N+1}) \cap L^\infty \hookrightarrow L^\infty$$

is compact. As a consequence of Proposition 1.7, we find that the inverse $(1+A)^{-1} : L^\infty \to L^\infty$ is compact.

This shows that A satisfies all requirements for applying the results obtained in §1 and §2 of this chapter.

- **Gradient inequalities**

As a direct consequence of Theorem 1.1 of §1, we have

THEOREM 3.1. **(Gradient inequalities)**

(i) Let $\varphi \in V \cap L^\infty$ be a critical point of $\mathcal{E}_F$. For any $\rho > 0$ there exist positive constants $c \equiv c(\rho), \sigma \equiv \sigma(\rho)$ depending on ρ, and $\theta \in [1/2, 1)$ that is independent from ρ, such that

$$c \cdot |\mathcal{E}_F(u) - \mathcal{E}_F(\varphi)|^\theta \leq \| - \alpha \Delta u + f(\cdot, u)\|_{W^{-1,2}} \tag{4.52}$$

for all $u \in V \cap L^\infty$ with $|\mathcal{E}_F(u) - \mathcal{E}_F(\varphi)| < \rho$ and $\|u - \varphi\|_{L^\infty} < \sigma$.

(ii) Let K be a bounded subset of $V \cap L^\infty$ that is contained in a compact subset of L^∞. Then there exists some function $\phi \in \mathcal{G}_2$ such that

$$\phi(\mathcal{E}_F(u)) \leq \| - \alpha \Delta u + f(\cdot, u)\|_{W^{-1,2}} \quad \forall u \in K.$$

In the above, $\alpha = diag(\alpha_1, ..., \alpha_d) > 0$ and $f \equiv \nabla_u F$.

REMARK 3.2. For the scalar case ($d = 1$), an "optimal" gradient inequality of the form (4.52) has been given in [**40**, Proposition 6.1], which is valid for all $u \in V \cap L^\infty$ with $|\mathcal{E}_F(u) - \mathcal{E}_F(\varphi)| < \rho$ and $\|u - \varphi\|_{L^2} < \sigma$. Certainly, a gradient inequality involving only the L^2 norm in the constraint conditions is the best possible one. However, some points of the proof in [**40**] are not clear. It is worth noting that those gaps do not destroy the truth of the main convergence result there [**40**, Theorem 3.1]. In fact, a careful check of the proof there reveals that a gradient inequality of our above form (4.52) involving the L^∞ norm in the constraint condition suffices. We will turn to this point in §5.

• **Convergence results**

As a special case of Theorem 1.2 of this chapter, under the choice of the triplet

$$(H, X, V) = (L^2, D(\sqrt{-\Delta}) \cap L^\infty, W^{1,2}),$$

we have the following convergence results. Here and below, we will use $\mathbf{B}_j u = 0$ ($j = 1, 2, 3, 4$) to represent that the function u satisfies one of the above listed boundary conditions **(B1)-(B4)**.

THEOREM 3.3. **(General form)**

Let the nonlinearity $f \equiv \nabla_u F$ be uniformly analytic in u.

(i) Assume that u is a global, bounded classical solution of the semilinear equation of heat type

$$\begin{cases} \dot{u} = J(\alpha\Delta u - f(x, u)) + g(t, u) \quad on \quad \Omega \times (0, \infty), \\ \mathbf{B}_j u = 0 \quad on \quad \partial\Omega \times (0, \infty), \end{cases} \tag{4.53a}$$

where

$$J : D(J) \subset L^2 \to L^2$$

is a closed, surjective and coercive linear operator, and the perturbation $g : \mathbb{R}_+ \times L^2 \to L^2$ satisfies the smallness condition

$$\begin{cases} \|g(t, u)\|_{L^2} \leq g(t) \quad \forall (t, u) \in \mathbb{R}_+ \times L^2, \\ \displaystyle\int_0^\infty \rho(t) g(t)^2 \, dt + \int_0^\infty \frac{1}{\rho(t)} \, dt < \infty \\ \text{with some increasing positive function} \quad \rho. \end{cases} \tag{4.53b}$$

If the orbit $O(u) = \{u(t) : t \geq 0\}$ is precompact in $W^{1,2} \cap L^\infty$, then there exists $\varphi \in W^{1,2} \cap L^\infty$ such that

$$\lim_{t\to\infty} \|u(t) - \varphi\|_{W^{1,2}\cap L^\infty} = 0, \quad \alpha\Delta\varphi = f(x, \varphi).$$

(ii) Assume that u is a global, bounded classical solution of the second order semilinear wave equation

$$\begin{cases} \ddot{u} + G(\dot{u}) = \alpha\Delta u - f(x, u) + g(t, u, \dot{u}) \quad on \quad \Omega \times (0, \infty), \\ \mathbf{B}_j u = 0 \quad on \quad \partial\Omega \times (0, \infty) \end{cases} \tag{4.54a}$$

where the damping map $G : D(G) \subset L^2 \to L^2$ is such that

$$\langle G(v), v\rangle_{L^2} \geq c \cdot (\|v\|^2 + \|G(v)\|^2_{W^{-1,2}}) \quad \forall v \in D(G) \tag{4.54b}$$

with a constant $c > 0$. Moreover, the perturbation $g : \mathbb{R}_+ \times L^2 \times L^2 \to L^2$ satisfies the smallness condition

$$\sup_{u,v\in L^2} \|g(\cdot, u, v)\|_{L^2} \leq \tilde{g}(\cdot), \qquad \int_0^\infty \left(\int_t^\infty \tilde{g}(s)^2 \, ds\right)^{1/2} dt < \infty. \tag{4.54c}$$

If

$$\{\dot{u}(t) : t \geq 0\} \subset W^{1,2} \cap L^\infty \cap D(G), \quad \sup\{\|\dot{u}(t)\|_{L^2} : t \geq 0\} < \infty,$$

and the orbit $O(u) = \{u(t) : t \geq 0\}$ *is precompact in* $W^{1,2} \cap L^\infty$, *then there exists some* $\varphi \in V \cap L^\infty$ *such that*

$$\lim_{t\to\infty} (\|\dot{u}(t)\|_{L^2} + \|u(t) - \varphi\|_{W^{1,2}\cap L^\infty}) = 0, \quad \alpha\Delta\varphi = f(x, \varphi).$$

PROOF. Case (i) corresponds to the first-order equation (3.28a) under the following choices:

$$H := L^2, \quad X := D(\sqrt{-\Delta}) \cap L^\infty, \quad \mathcal{N} := J\mathcal{E}'_F.$$

Thus, our convergence result for case (i) follows from an application of Theorem 1.2 (i) of this chapter.

Case (ii) corresponds to the second-order equation (3.32a) with the negative sign choice, $\kappa := 1$, and the following choices of the spaces and the gradient map:

$$H := L^2, \quad X := D(\sqrt{-\Delta}) \cap L^\infty, \quad \mathcal{E}' := \mathcal{E}'_F.$$

Moreover, the map B is chosen as $B(t, u, v) := Gv$. Note that the coercivity condition in (3.32b) for the map B is ensured by the coercivity $\langle G(v), v\rangle \geq c^2\|v\|^2$ $(v \in D(G))$ given in (4.54b) above. Furthermore, the inequality $c \cdot \|G(v)\|^2_{W^{-1,2}} \leq \langle G(v), v\rangle_{L^2}$ $(v \in D(G))$ given in (4.54b) ensures that the map $B(t, u, v) := Gv$ satisfies condition (3.38b) with $V := W^{1,2}$ and the constant $c_1 := 1/c$. Therefore, our convergence result for case (ii) follows from an application of Theorem 1.2 (ii) of this chapter. □

We make several remarks to Theorem 3.3 (ii), which applies to semilinear wave equations.

(a) Theorem 3.3 (ii) applies particularly to the special choice $G(v) := \beta v$ (with $\beta > 0$), which corresponds to the damped semilinear wave equations $\ddot{u} + \beta\dot{u} = \alpha\Delta u - f(x, u) + g(t, u, \dot{u})$ with a linear damping. For this special case, the convergence result given in Theorem 3.3 (ii) has been obtained previously by Jendoubi, Hauraux, Chill *et al.* [**71, 21, 59, 61**].

(b) Chill and Fašangová [**19**] have obtained the same convergence result as Theorem 3.3 (ii) for the case where G is a linear damping that might be unbounded. We point out that their results also apply to a more general linear damping of the form $B_0\dot{u} + a * B_1\dot{u}$ for which the semilinear wave equations are kinds of integral equations. Note that our Theorem 4.5 in Chapter 3 is applicable to such a case.

(c) The special choice $G(v) := \beta(-\Delta_D)v$ (with a positive constant β) in (4.54a), under the Dirichlet boundary condition, yields the following system of semilinear wave equations:

$$(*) \qquad \begin{cases} \ddot{u} - \beta\Delta\dot{u} = \alpha\Delta u - f(x, u) + g(t, u, \dot{u}) & \text{on} \quad \Omega, \\ u = 0 \quad \text{on} \quad \partial\Omega. \end{cases}$$

Such a system arises in the study of nonlinear viscoelasticity (see Prüss [**99**] for more details), and $u = (u_1, u_2, ..., u_d)$ describes the displacement of a collection of d viscoelastic bodies on the bounded domain $\Omega \subset \mathbb{R}^N$. Our above Theorem 3.3 (ii) asserts that if $f = \nabla_u F$ is analytic, then every bounded and compact global solution of $(*)$ converges to one single equilibrium (cf. Chill-Fašangová [**19**, Theorem 4.6]).

Next we turn to a special case of the first-order system (4.53a). We take $J := Id_H$ and $g(t, u) \equiv 0$ in (4.53a). The resulting system is the following semilinear gradient-like Reaction-Diffusion system

$$(4.53a') \qquad \begin{cases} \dot{u} = \alpha \Delta u - f(x, u) & \text{on} \quad \Omega \times (0, \infty), \\ \mathbf{B}_j u = 0 \quad \text{on} \quad \partial\Omega \times (0, \infty). \end{cases}$$

In order to apply the results in §1 of this chapter to (4.53a′), we assume the following growth condition of the form (4.18c) for the nonlinearity: There exist two vectors $\tau_+ = (\tau_+^1, \tau_+^2, ..., \tau_+^d) \geq 0$ and $\tau_- = (\tau_-^1, \tau_-^2, ..., \tau_-^d) \leq 0$ such that for all boundary points $u \equiv (u_1, ..., u_d) \in \partial Q_\tau$ of the rectangle $Q_\tau := \{u \in \mathbb{R}^d : \tau_- \leq u \leq \tau_+\}$, there hold the inequalities:

$$(4.53b') \qquad \begin{cases} f_j(\cdot, u) \geq 0 & \text{if} \quad u_j = \tau_+^j; \\ f_j(\cdot, u) \leq 0 & \text{if} \quad u_j = \tau_-^j; \end{cases} \qquad j = 1, 2, ..., d.$$

Here, f_j's are the components of $f \equiv \nabla_u F$.

THEOREM 3.4. **(Special form)**
Assume that the nonlinearity $f \equiv \nabla_u F$ in (4.53a′) is uniformly analytic in u with $f(\cdot, 0) \in L^\infty$, and satisfies the growth condition (4.53b′). Set $X := D(\sqrt{-\Delta}) \cap L^\infty$ and $\mathcal{S}_\tau := \{\varphi \in X : \mathcal{E}_F'(\varphi) = 0, \tau_- \leq u \leq \tau_+\}$. Then for any initial value $u_0 \in L^\infty$ with $\tau_- \leq u_0 \leq \tau_+$, the Cauchy problem for (4.53a′) has a unique classical solution

$$u \in C([0, \infty); L^2) \cap C((0, \infty); D(\Delta)) \cap C^1((0, \infty); L^2)$$

for which there exists an equilibrium $\varphi \in \mathcal{S}_\tau$ such that

$$\lim_{t \to \infty} \|u(t) - \varphi\|_{W^{1,2} \cap L^\infty} = 0.$$

PROOF. The operator $A := -\alpha\Delta$ is diagonal; thus, Theorem 1.4 applies to (4.53a′) and yields that the interval $[\tau_-, \tau_+]$ is an invariant domain for (4.53a′), and the orbit $O(u)$ of every solution u of (4.53a′) with an initial value $u_0 \in [\tau_-, \tau_+]$ is precompact in $W^{1,2} \cap L^\infty$. Hence, the convergence result follows from Theorem 1.8. □

As direct consequences of Proposition 1.5 and Theorem 1.9 of this chapter, we have the following existence and stability result for local ground states concerning the gradient system (4.53a′).

PROPOSITION 3.5. **(Stability of local ground states; solutions of semilinear elliptic systems of gradient type)**

Assume that the nonlinearity $f \equiv \nabla_u F$ in (4.53a′) is locally Lipschitz continuous in u, uniformly for $x \in \Omega$, and such that $f(\cdot, 0) \in L^\infty$ and satisfies condition (4.53b′). Then there hold the following assertions (i)–(ii):

(i) The subset

$$\mathcal{S}_\tau := \{\varphi \in X : \mathcal{E}_F'(\varphi) = 0,\ \tau_- \leq \varphi \leq \tau_+\}$$

of solutions of the semilinear elliptic system

$$(*) \qquad \begin{cases} \alpha\Delta u = f(x,u), & \tau_- \leq u \leq \tau_+ \\ \mathbf{B}_j u = 0 & on \quad \partial\Omega \end{cases}$$

is non-empty and compact in $X = D(\sqrt{-\Delta}) \cap L^\infty$.

(ii) There exists some $\varphi \in \mathcal{S}_\tau$ such that

$$\mathcal{E}_F(\varphi) \leq \mathcal{E}_F(u) \quad \forall u \in X \quad with \quad \tau_- \leq u \leq \tau_+.$$

Moreover, if $f(x,u)$ is uniformly analytic in u, then φ is Lyapunov stable in the sense that for any given $\varepsilon > 0$, there exists a constant $\delta \equiv \delta(\varepsilon) > 0$ with the following property: For any initial value $u_0 \in L^\infty$, $\tau_- \leq u_0 \leq \tau_+$, with

$$\|u_0 - \varphi\|_{L^2} < \delta,$$

there holds

$$\sup_{t\geq 1} \|u(t) - \varphi\|_{W^{1,2}\cap L^\infty} < \varepsilon,$$

where u is the unique, global and classical solution of the gradient system (4.53a′) with $u(0) = u_0$.

Furthermore, if φ is an isolated critical point of $\mathcal{E}_F$ in $\mathcal{S}_\tau$, then φ is uniformly asymptotically stable in the sense that there exists a constant $\delta_0 > 0$ such that

$$\|u(t) - \varphi\|_{W^{1,2}\cap L^\infty} \to 0 \quad \text{as} \quad t \to \infty$$

uniformly for all solutions u of (4.53a′) with initial data $u_0 \in L^\infty$ satisfying $\tau_- \leq u_0 \leq \tau_+$ and $\|u_0 - \varphi\|_{L^2} < \delta_0$.

REMARK 3.6. There is a vast literature on studying the above semilinear elliptic problem $(*)$ (mainly with $d = 1$); see, e.g., [**16, 31, 112, 122**] and the references therein.

3.2. Perfect nonlinearity: $N = 2, 3$. We consider the case $N = 2, 3$, i.e., Ω is either a planar or a three-dimensional domain. In order to simplify the notation, we consider only the scalar case, i.e., we set

$$d = 1, \quad \alpha_1 = 1.$$

The general vectorial case $d > 1$ can be treated similarly.

Since $d = 1$, we have

$$V = D(\sqrt{-\Delta}) = \{u \in W^{1,2}(\Omega) : \mathbf{B}_j u = 0\}.$$

Consider the functional

$$\mathcal{F}(u) := \int_\Omega F(x,u)\,dx \quad (u \in V).$$

We want to apply results (Theorems 2.3 and 2.4) obtained in §2 for the perfect case. To do so, we need to find conditions on the analytic function F such that both composition operators

$$V \ni u \mapsto F(x,u) \in L^1, \quad V \ni u \mapsto f(x,u) \in L^2 \tag{4.55a}$$

are well-defined and analytic, and such that

$$\{u \in V : \mathcal{E}_F(u) \leq R\} \quad \text{is bounded in} \quad V \tag{4.55b}$$

for every $R > 0$, where

$$\mathcal{E}_F(u) := \frac{1}{2}\int_\Omega |\nabla u|^2\,dx + \int_\Omega F(x,u)\,dx. \tag{4.55c}$$

In (4.55a), $f \equiv \partial_u F$. Note that an easy condition that ensures the boundedness in (4.55b) is the following coercivity condition

$$F(\cdot,u) \geq c_1 u^2 - c_2 \quad \forall u \in \mathbb{R} \tag{4.55d}$$

with two positive constants c_1, c_2. We treat the cases $N = 2$ and $N = 3$ separately as follows. First we derive the conditions on F corresponding to the requirements in (4.55) and then assemble the related results as applications of Theorem 2.3 and Corollary 2.4.

• $N = 3$: **Fourth-order polynomial nonlinearity is the perfect choice for three-dimensional domains.**

By the Sobolev Embedding Theorem we have

$$V \subset W^{1,2}(\Omega) \hookrightarrow L^6(\Omega) \tag{4.56a}$$

and the embedding $W^{1,2}(\Omega) \hookrightarrow L^p(\Omega)$ is compact if $p < 6$; see Gilbarg-Trudinger [**45**, Theorem 7.26, p. 171]. According to the analyticity requirement for the composition operators given in (4.55a) and the coercivity, we see that F must be a polynomial of order less than 5, i.e., F has the representation

$$F(x,u) = \sum_{j=0}^{4} a_j(x)u^j \quad \text{with} \quad a_j \in L^\infty. \tag{4.56b}$$

Correspondingly,

$$f(x,u) = \partial_u F(x,u) = \sum_{j=0}^{3}(j+1)a_{j+1}(x)u^j. \tag{4.56c}$$

Clearly, condition (4.55d) will be satisfied if the leading coefficient a_4 is strictly positive, i.e., if

$$a_4(\cdot) \geq c_0 > 0. \tag{4.56d}$$

In short, we have shown that both Theorems 2.3 and 2.4 of this chapter apply to the case where $N = 3$ and the fourth-order polynomial nonlinearity F satisfies (4.56d).

• $N = 2$: **Analytic nonlinearity of sub-quadratic exponential type is the perfect choice for planar domains.**

For the sake of simplicity, we treat only the case of Dirichlet boundary condition. The other boundary conditions can be handled similarly.

We prepare some estimates. For $u \in W_0^{1,1}(\Omega)$ we have (see Eq. (7.37) in Gilbarg-Trudinger [**45**, p. 161]) that

$$|u| \leq \frac{1}{2\pi} V_{\frac{1}{2}} |\nabla u|, \tag{4.57a}$$

where the operator $V_{\frac{1}{2}}$ defined by the Riesz potential

$$(V_{\frac{1}{2}} u)(x) := \int_\Omega \frac{u(y)}{|x-y|}\, dy \quad (u \in L^1(\Omega))$$

obeys the estimates:

$$\|V_{\frac{1}{2}} u\|_{L^q} \leq \sqrt{\pi} |\Omega|^{1/q} \left(\frac{q}{2} + 1\right)^{\frac{1}{2}+\frac{1}{q}} \cdot \|u\|_{L^2} \tag{4.57b}$$

for all $u \in L^2(\Omega)$ and $q \geq 1$, cf. Gilbarg-Trudinger [**45**, Lemma 7.12, p. 159]. Taking (4.57a,b) together, we find that

$$\|u\|_{L^q} \leq \frac{1}{2\sqrt{\pi}} |\Omega|^{1/q} \left(\frac{q}{2} + 1\right)^{\frac{1}{2}+\frac{1}{q}} \cdot \|\nabla u\|_{L^2} \quad \forall u \in W_0^{1,2}(\Omega) = V. \tag{4.57c}$$

Fix r, $2 \leq r \leq \infty$. We consider a function $F : \Omega \times \mathbb{C} \to \mathbb{C}$ that satisfies the coercivity condition (4.55d) and admits the following representation:

$$F(x, z) = \sum_{n=0}^{\infty} a_n(x) z^n, \tag{4.58a}$$

where $\{a_n\}$ is a sequence of **real** functions in $L^r(\Omega)$ and obeys the following growth condition

$$\forall \varepsilon > 0\, \exists C > 0 \quad \text{such that} \quad \sqrt{n!} \cdot \|a_n\|_{L^r} \leq C \cdot \varepsilon^n \quad \forall n = 0, 1, 2, \ldots. \tag{4.58b}$$

It follows from (4.58b) that for a.e. $x \in \Omega$, the function $F(x, z)$ is an entire function of $z \in \mathbb{C}$. Therefore, the maximum modulus function

$$M(x, R) := \sup\{|F(x, z)| : |z| = R\} \quad (R \geq 0) \tag{4.58c}$$

is a Carathéodory function and, for each $R \geq 0$, the function $M(\cdot, R) \in L^r(\Omega)$. An elementary approach shows that the growth condition (4.58b) for the coefficients $\{a_n\}$ is equivalent to the following growth condition on the maximum modulus

$$\lim_{R \to \infty} \frac{\ln \|M(\cdot, R)\|_r}{R^2} = 0. \tag{4.58d}$$

By virtue of (4.58d), one sees that the functions satisfying (4.58a,b) form a (real) subalgebra of the algebra of the functions over $\Omega \times \mathbb{C}$. Concrete examples of such functions are $F(x,u) \equiv e^u$ and $F(x,u) \equiv \sin u$. Moreover, (4.58d) reveals that analytic functions satisfying growth condition (4.58b) are nothing but these analytic functions with an exponential type less than 2, i.e., a sub-quadratic exponential type.

LEMMA 3.7. *Assume (4.58a,b). Then the composition operator* $\mathcal{K}$

$$\mathcal{K}(u)(x) := F(x, u(x))$$

maps $W_0^{1,2}(\Omega) = V$ *into* $\bigcap_{p<r} L^p(\Omega)$. *Moreover, for each* $p \in [2, r)$, *the map*

$$\mathcal{K} : W_0^{1,2}(\Omega) \to L^p(\Omega)$$

is analytic, and maps bounded subsets of $W_0^{1,2}(\Omega)$ *into bounded subsets of* $L^p(\Omega)$.

PROOF. Using (4.58a), we rewrite $\mathcal{K}(u)$ as

$$\mathcal{K}(u)(x) = \sum_{n=0}^{\infty} a_n(x) u(x)^n. \tag{4.59a}$$

Fix $p \in [2, r)$. For $u \in V$ we have, by Hölder's inequality, that

$$\|a_n u^n\|_{L^p} \leq \|a_n\|_{L^r} \cdot \|u\|_{L^{ns}}^n \tag{4.59b}$$

with $s := pr/(r-p)$. For $n \geq 1$ we use (4.57c) with $q = ns > p \geq 2$ to find that

$$\|u\|_{L^{ns}}^n \leq |\Omega|^{1/s} (ns)^{\frac{1}{s}+\frac{n}{2}} \cdot \|\nabla u\|_{L^2}^n.$$

Therefore, by (4.58b) we obtain for all $n \geq 1$ that

$$\|a_n u^n\|_{L^p} \leq C \cdot (ns|\Omega|)^{1/s} \cdot (\varepsilon \|\nabla u\|_{L^2})^n \cdot \left(\frac{n^n}{n!}\right)^{1/2},$$

which, by virtue of the Stirling estimate $n^n \leq n! e^n$, yields that

$$\|a_n u^n\|_{L^p} \leq C \cdot (ns|\Omega|)^{1/s} \cdot (\sqrt{e}\varepsilon \|\nabla u\|_{L^2})^n \quad n \geq 1. \tag{4.59c}$$

Since $\varepsilon > 0$ is arbitrary, (4.59c) implies that $\sum_{n=0}^{\infty} \|a_n u^n\|_{L^p} < \infty$ for all $u \in V$ and the convergence is uniform for u in any bounded subset of V. From this we obtain the inclusion $\mathcal{K}(V) \subset L^p(\Omega)$ and the analyticity as well as the boundedness of the map $\mathcal{K} : V \to L^p(\Omega)$ for all $p \in [2, r)$. This completes the proof. □

Applying Lemma 3.7 to the choice $p = 2 < r$ we claim that both composition operators given by (4.55a) are analytic, provided conditions (4.58a,b) are satisfied.

• **Convergence results for** $N = 2, 3$.

Summing up, in the above we have shown that the required conditions in (4.55a,b) are satisfied, provided the function F has the following properties:

$N = 3$: The function F is a fourth-order polynomial of the form

$$F(x,u) = \sum_{j=0}^{4} a_j(x) u^j \quad \text{with} \quad a_j \in L^\infty; \quad a_4 \geq c_0 > 0. \tag{4.60a}$$

$N = 2:$ The function F has the representation

$$F(x, z) = \sum_{n=0}^{\infty} a_n(x) z^n \tag{4.60b}$$

where $\{a_n\}$ is a sequence of **real** functions in $L^r(\Omega)$ with some $r \in [2, \infty]$ and obeys the following growth condition:

$$\forall \varepsilon > 0\, \exists C > 0 \quad \text{such that} \quad \sqrt{n!} \cdot \|a_n\|_{L^r} \le C \cdot \varepsilon^n \quad \forall n = 0, 1, 2, \dots. \tag{4.60c}$$

Moreover, there exist two positive constants c_1, c_2 such that

$$F(\cdot, u) \ge c_1 u^2 - c_2 \quad \forall u \in \mathbb{R}. \tag{4.60d}$$

With such a function F we consider the following gradient-like system

$$\begin{cases} \dot{u} = \Delta u - f(x, u) + g(t, u),\ u(0) = u_0, \\ \mathbf{B}_j u = 0 \quad \text{on} \quad \partial\Omega \end{cases} \tag{4.61a}$$

in a bounded domain $\Omega \subset \mathbb{R}^N$ with $N = 2$ or $N = 3$, where $f \equiv \partial_u F$ and the perturbation $g : \mathbb{R}_+ \times V \to L^2(\Omega)$ is a locally Lipschitz map satisfying

$$\begin{cases} \|g(t, u)\|_{L^2} \le g(t) \quad \forall (t, u) \in \mathbb{R}_+ \times V, \\ g \in L^\infty, \quad \displaystyle\int_0^\infty \rho(t) g(t)^2\, dt + \int_0^\infty \frac{1}{\rho(t)}\, dt < \infty \\ \text{with some increasing positive function} \quad \rho. \end{cases} \tag{4.61b}$$

Correspondingly, we have the gradient system

$$\begin{cases} \dot{u} = \Delta u - f(x, u), \\ \mathbf{B}_j u = 0 \quad \text{on} \quad \partial\Omega. \end{cases} \tag{4.61c}$$

THEOREM 3.8. **(Convergence; stability of ground state)**
Under the above circumstances we have the following assertions (i)–(ii):

(i) For any initial value $u_0 \in V = D(\sqrt{-\Delta})$, the Cauchy problem (4.61a,b) has a unique global classical solution

$$u \in C([0, \infty); V) \cap C(0, \infty); D(\Delta)) \cap C^1((0, \infty); L^2)$$

such that

$$\lim_{t \to \infty} \|u(t) - \psi\|_{W^{1,2}} = 0$$

with some equilibrium $\psi \in \mathcal{S}$, where $\mathcal{S}$ is the set of all weak solutions of the coercive semilinear elliptic problem $\Delta u = f(x, u)$ on Ω, $\mathbf{B}_j u = 0$ on $\partial\Omega$.

(ii) There exists some $\varphi \in \mathcal{S}$ such that

$$\mathcal{E}_F(\varphi) \le \mathcal{E}_F(u) \quad \forall u \in V.$$

Moreover, this ground state φ is Lyapunov stable in the sense that for any given $R > \mathcal{E}_F(\varphi)$ and $\varepsilon > 0$, there exists a constant $\delta \equiv \delta(R, \varepsilon) > 0$ with the following property: For any initial value $u_0 \in V$ with

$$\mathcal{E}_F(u_0) \le R, \quad \|u_0 - \varphi\|_{L^2} < \delta,$$

there holds

$$\sup_{t\geq 1} \|u(t) - \varphi\|_{W^{1,2}} < \varepsilon,$$

where u is the unique, global and classical solution of the gradient system (4.61c) with the initial value $u(0) = u_0$.

Furthermore, if φ is an isolated critical point of $\mathcal{E}_F$, then φ is uniformly asymptotically stable in the sense that for each $R > \mathcal{E}_F(\varphi)$ there exists some constant $\delta_0 > 0$ such that

$$\|u(t) - \varphi\|_{W^{1,2}} \to 0 \quad \text{as} \quad t \to \infty$$

uniformly for all solutions u of the gradient system (4.61c) with initial data $u_0 \in V$ satisfying $\mathcal{E}_F(u_0) \leq R$ and $\|u_0 - \varphi\|_{L^2} < \delta_0$.

PROOF. (i) is a consequence of Theorem 2.3 and (ii) is a consequence of Theorem 2.4. □

Roughly speaking, the convergence result in Theorem 3.8 says that all classical global solutions to the coercive semilinear heat equation $\dot{u} = \Delta u - f(x,u)$ (plus one of the boundary conditions **(B1)-(B4)**) on a bounded domain $\Omega \subset \mathbb{R}^N$ will converge as $t \to \infty$ to one single equilibrium provided that the nonlinearity f is a polynomial of order three with a positive leading coefficient if $N = 3$ and is an analytic function in u of sub-quadratic exponential type if $N = 2$. Moreover, this convergence property is stable under bounded perturbations g that satisfy the smallness condition (4.61b). We remark that there holds the convergence to one single equilibrium if $N = 1$ or if Ω is a thin domain (see Hale-Raugel [**53**]).

Finally, we mention that the corresponding gradient inequality for the above special perfect cases (as a direct consequence of Prop. 2.2) has the following form: For any $R > 0$ there exists some $\phi \in \mathcal{G}_2$ such that

$$\phi(\mathcal{E}_F(u)) \leq \|\Delta u - f(x,u)\|_{W^{-1,2}} \quad \forall u \in V, \|u\|_{W^{1,2}} \leq R. \tag{4.62}$$

Certainly, a gradient inequality of the form (4.62) with the natural definition domain of the functional is the best one.

3.3. Swift-Hohenberg equations: a perfect case for $N \leq 3$. The Swift-Hohenberg (SH) equations [**114, 98, 11, 26**] are fourth-order gradient systems of the form

$$\dot{u} = -(\Delta + q_0^2)^2 u - f(u) \tag{SH}$$

associated with the energy functional

$$\mathcal{E}_F(u) := \frac{1}{2}\int_\Omega [(\Delta + q_0^2)u]^2\, dx + \int_\Omega F(u)\, dx \quad (F' = f), \tag{4.63a}$$

where q_0 is a real constant. The corresponding complex generalization [**11**] has the form

$$\dot{\psi} = -(\Delta + q_0^2)^2 \psi - f(|\psi|^2)\psi, \tag{CSH}$$

(4.63b) $$\mathcal{E}_F(\psi) := \frac{1}{2}\int_\Omega |(\Delta + q_0^2)\psi|^2\,dx + \frac{1}{2}\int_\Omega F(|\psi|^2)\,dx \quad (F' = f),$$

where f is a real valued function on $\mathbb{R}$.

The original (SH) equation posed by J. Swift and P. C. Hohenberg [**114**] corresponds to the choice of the nonlinearity f :

(4.64) $$f(u) = \frac{1}{4}u^4 - \frac{1}{3}\beta u^3 + \frac{1}{2}\alpha u^2$$

with physical parameters α, β. This is a model for the crystallization of a fluid in equilibrium. The importance of this model is, as shown by Swift and Hohenberg [**114**], that Langevin equations for hydrodynamic fluctuations are equivalent, near the instability, to the original (SH) model. Other aspects of the (SH) model are studied numerically in [**98**] (concerning the wavelength selection in cellular flows) and in [**11**] (concerning the stable localized structure of two- and three-dimensional fields for both Eq. (SH) and Eq. (CSH)).

We consider the physically reasonable case $N \leq 3$. We will show that if the nonlinearity F is bounded below and analytic, then any "good" initial value develops a global solution to Eq. (SH) (resp., Eq. (CSH)) which converges to some equilibrium. In particular, the above-mentioned fourth-order polynomial nonlinearity considered by Swift and Hohenberg ensures this convergence property.

To confirm this, we assume that $F : \mathbb{R} \to \mathbb{R}$ is a (real) analytic function that is bounded below. Set

(4.65a) $$X := D(\Delta) \subset W^{2,2}(\Omega),$$

where the Laplacian Δ is assigned with one of the classical boundary conditions **(B1)**–**(B4)**. Consider the following symmetric and positive definite bilinear form on X :

(4.65b) $$a(u,v) := \langle(\Delta + q_0)u, (\Delta + q_0)v\rangle_{L^2} + \langle u, v\rangle_{L^2} \quad (u, v \in X).$$

It follows from the Lax-Milgram Theorem [**45**, Theorem 5.8, p. 83] that there exists a self-adjoint and positive definite linear operator

$$B : D(B) \subset L^2 \to L^2$$

such that

(4.65c) $$\begin{cases} D(B) = X = D(\Delta), \\ \langle Bu, Bv\rangle_{L^2} = a(u,v) \quad \forall u, v \in X. \end{cases}$$

By the Sobolev Embedding Theorem we have

$$D(B) = D(\Delta) \subset W^{2,2}(\Omega) \hookrightarrow C^\alpha(\bar{\Omega})$$

for all $0 \leq \alpha < \min\{1, 2 - N/2\}$ and the embeddings are compact. The positivity $2 - N/2 > 0$ is ensured by the assumption $N \leq 3$. In particular, we see that the embedding

$$X = D(B) \hookrightarrow C(\bar{\Omega})$$

is compact. Therefore, by the analyticity of F we find that both composition operators

$$X \ni u \mapsto F(u) \in C(\bar{\Omega}), \quad X \ni u \mapsto f(u) \in C(\bar{\Omega})$$

are compact and analytic. Thus, our condition (4.33a) corresponding to the choice $\mathcal{F}(u) := \int_\Omega F(u)\,dx$ is satisfied and hence both Theorems 2.3 and 2.4 of this chapter apply. It yields the following results (i)–(ii):

(i) For every initial value $u_0 \in D(B) = D(\Delta)$, the Cauchy problem corresponding to Eq. (SH) has a classical and global solution

$$u \in C([0,\infty); D(\Delta)) \cap C((0,\infty); D(\Delta^2)) \cap C^1((0,\infty); L^2)$$

for which there exists an equilibrium $\varphi \in D(\Delta)$ of Eq. (SH) such that

$$\lim_{t\to\infty} \|u(t) - \varphi\|_{W^{2,2}} = 0.$$

More importantly, our Theorem 2.3 says that this convergence is stable under deviations g that satisfy a smallness condition of the form (4.61b). From the point of view of the physical applications of these models, this stability is certainly very important.

(ii) The set of equilibria of Eq. (SH) is a compact subset of $D(\Delta)$. Moreover, the energy functional $\mathcal{E}_F$ attains its absolute minimum, i.e., there exists some $\xi \in D(\Delta)$ such that

$$\mathcal{E}_F(\xi) \leq \mathcal{E}_F(u) \quad \forall u \in D(\Delta).$$

Moreover, this ground state ξ is Lyapunov stable (resp., uniformly asymptotically stable) in the sense given by Theorem 2.4.

Similar assertions hold true for the complex model Eq. (CSH). Here we omit the details.

3.4. Convergence in the Cahn-Hilliard model. The Cahn-Hilliard (CH) model (see Temam [**118**]) was proposed as a continuum model for the description of the dynamics of pattern formation in phase transition. It is a fourth-order system of the form:

$$(CH) \qquad \begin{cases} \dot{u} = \Delta(-\Delta u + f(x,u)) + g(t,u) \quad \text{on} \quad \Omega, \\ u = 0 \quad \text{on} \quad \partial\Omega. \end{cases}$$

The autonomous part (obtained by switching g into zero) is not a gradient system. However, the equation (CH) is in the form of (4.53a) corresponding to the special choice $J = -\Delta$, which is coercive under the Dirichlet boundary condition, by virtue of the generalized Poincaré inequality (see Temam [**118**] or Maz'ja [**87**]). Furthermore, when the function f is analytic and the perturbation g satisfies the smallness condition (4.53b), then Theorem 3.3 (i) of this chapter applies and yields the convergence as $t \to \infty$ to one single equilibrium of a compact solution of the perturbed (CH) equation.

The above-mentioned stability of convergence of compact solutions of the (CH) equation under "small" perturbations g (which satisfies the smallness condition (4.53b)) is, from the point of view of the physical applications of the (CH) model, very important.

In [**102**], Rybka and Hoffmann have treated the unperturbed Cahn-Hilliard equation (i.e., $g \equiv 0$) for the case $N = 1, 2, 3$. They showed that if the growth of f satisfies $|f(x, u)| \leq a + b|u|^r$ for some positive constants a, b and r, $r \in [2, \infty)$ if $N = 1, 2$, or $r \in [2, 4]$ if $N = 3$, then every initial value $u_0 \in D(\Delta)$ develops a classical global solution of the (CH) equation whose orbit is contained in a compact subset of $C^4(\bar{\Omega})$ and thus converges as $t \to \infty$ to one single equilibrium in the $C^4(\bar{\Omega})$ norm topology.

The original (CH) equation corresponds to the choice

$$f(x, u) \equiv -\alpha u + \beta u^3,$$

where $\alpha, \beta > 0$ are physical parameters. In [**118**], Temam treated the following general choices of f :

$$f(x, u) \equiv f(u) = \sum_{j=1}^{2r-1} a_j u^j \quad (r \in \mathbb{N}, r \geq 2). \tag{4.66}$$

Under the assumption that the leading coefficient of f is positive, i.e., $a_{2r-1} > 0$, he establishes the existence of compact attractors for the corresponding unperturbed (CH) equation. The positivity $a_{2r-1} > 0$ is used to guarantee that the energy functional

$$\mathcal{E}_F(u) := \frac{1}{2}\|\nabla u\|_{L^2}^2 + \int_\Omega F(u)\, dx \quad (F' = f)$$

of the solutions is bounded from below. This boundedness of the energy combined with the growth conditions like the ones used by Rybka and Hoffmann is enough to deduce the uniform $C^k(\bar{\Omega})$ boundedness of the solutions, and thus the convergence of the solutions to equilibria.

3.5. A remark. The results and proofs given above remain valid if the Laplacian operator Δ is replaced by a general symmetric strongly elliptic operator

$$\mathcal{A} = \sum_{i,j}^{N} D_i(a_{ij}(x) D_j) \qquad (D_k := \frac{\partial}{\partial x_k}).$$

The submarkovian property of $-\mathcal{A}$, i.e., the property that $\mathcal{A}$ generates an analytic semigroup of contractions on every L^p space with $1 < p < \infty$ is well known, see, e.g., Pazy [**94**, Theorem 3.6]. Note that there also holds the submarkovian property and compactness property for the strongly elliptic operators on compact Riemannian manifolds. Hence, all of the above results and proofs remain valid for strongly elliptic operators on compact Riemannian manifolds.

3.6. A quick discussion: Why does the convergence fail? This problem is primarily concerned with the semilinear heat equation

$$\dot{u} = \Delta u - f(x,u) \quad \text{on} \quad \Omega, \quad u|_{\partial\Omega} = 0,$$

where the Dirichlet boundary condition is considered. Poláčik and Simondon [**96**] show that for any given bounded domain $\Omega \subset \mathbb{R}^N$ with $N \geq 2$, there exists a smooth function $f \in C^\infty(\mathbb{R}^{N+1})$ such that the above heat equation has a bounded solution whose ω-limit set is a continuum of equilibria. The functions $f(x,u)$ constructed by Poláčik and Simondon are dependent on the space variable $x \in \Omega$. Thus, it remains a problem as to whether there exists a smooth function $f(x,u) \equiv f(u)$ with the same nonconvergent property but that is independent of x.

Concerning the above problem, we mention that the geometry of the domain Ω can help a lot in proving convergence. For example, it was shown by Haraux and Poláčik [**62**] that if Ω is a ball, then every positive solution to the above heat equation with a space-independent nonlinearity $f(x,u) = f(u) \in C^2(\mathbb{R})$ will converge as $t \to \infty$ to one single equilibrium. Their proof relies on the fact that every positive equilibrium for this case is spherically symmetric.

In the unbounded case $\Omega = \mathbb{R}^N$, Poláčik and Yanagida [**97**] find that nonconvergent compact solutions exist in the following semilinear heat equation

$$(*)_p \qquad \dot{u} = \Delta u + |u|^{p-1}u \quad \text{in} \quad \mathbb{R}^N,$$

where $N \geq 11$ and $p \geq p_c := (N - 2\sqrt{N-1})/(N - 4 - 2\sqrt{N-1})$. The nonconvergent solution constructed by them is located between two positive equilibria $\varphi_\alpha \leq \varphi_\beta$ of $(*)_p$ (both $\varphi_{\alpha,\beta}$ are in $C_0(\mathbb{R})$) and oscillates in such a way that its ω-limit set forms a monotone arc of equilibria between the two positive equilibria φ_α and φ_β.

It is interesting to note that the nonlinearity $f(u) := |u|^{p-1}u$ will be *analytic* if the exponent p ($p \geq p_c$) is an odd integer. This reveals that analyticity of the nonlinearity does not help at all for this case. From this we see that for the unbounded case $\Omega = \mathbb{R}^N$ ($N \geq 11$), the analytic gradient map $\mathcal{E}_p'(u) := \Delta u + u^p$ ($p \in \mathbb{N}$ is odd and $p \geq p_c$) does not satisfy a gradient inequality near the equilibria $\varphi_{\alpha,\beta}$. It seems that the reason for the invalidity of gradient inequalities is that the linearization $\mathcal{E}_p''(\varphi) = \Delta + p\varphi^{p-1}\bullet$ of the analytic gradient map $\mathcal{E}_p'$ around the equilibria $\varphi = \varphi_{\alpha,\beta} \in C_0(\mathbb{R})$, when it acts in the natural Hilbert space $L^2(\mathbb{R})$, is not a Fredholm operator of index zero. However, we see that the Fredholm property of the linearization $\mathcal{E}_p''(\varphi)$ and thus the validity of gradient inequality can be guaranteed if the underlined space is made appropriately small – one such example is to use some weighted spaces. This observation helps us to understand the convergence result of Poláčik and Yanagida [**97**, Theorem 1.1] (see also Gui, Ni and Wang [**48, 49**]), which says that there holds the convergence $\|u(t) - \varphi_\alpha\|_{L^\infty} \to 0$ as $t \to \infty$ if the solution u of $(*)_p$ has an initial value that locates closely enough, in some appropriate sense, to the equilibrium φ_α.

Remember that in the bounded case, the convergence to one single equilibrium will be ensured if the nonlinearity is analytic – the main point for proving this convergence result is the validity of a gradient inequality of LS-type near every equilibrium.

4. Ginzburg-Landau models for superconductivity

These models have been studied in [**17, 116, 117, 78, 41, 70**], and the convergence of solutions to equilibria has been established just recently (2001) by Feireisl and Takáč [**41**] (see also Huang and Takáč [**70**]). Lin and Du [**78**] have obtained the convergence result under very restrictive conditions; however, as pointed out in [**41**], the proof of Lin and Du is somewhat incomplete.

Below we will show how our abstract Theorem 4.1 in §4 of Chapter 3 can be applied to rederive the convergence result of solutions to the time-dependent Ginzburg-Landau (TDGL) equations of superconductivity [**46**].

For simplicity, we consider only the (TDGL) equations in the *zero electric potential gauge*. Then the (TDGL) equations governing the complex-valued *order parameter* ψ and the real vector-valued *vector potential* $\mathbf{A}$ have the form:

$$\begin{cases} \eta\frac{\partial\psi}{\partial t} = -\left(\frac{i}{\kappa}\nabla + \mathbf{A}\right)^2\psi + (1-|\psi|^2)\psi, \\ \frac{\partial \mathbf{A}}{\partial t} = -\nabla\times\nabla\times\mathbf{A} + \mathbf{J}_s(\psi,\mathbf{A}) + \nabla\times\mathbf{H} \end{cases} \quad \text{in} \quad \Omega. \tag{4.67a}$$

Here $\Omega \subset \mathbb{R}^N$ ($N = 2$ or $N = 3$) is the region occupied by the superconducting material. We assume that Ω is a bounded domain with a boundary $\partial\Omega$ of class $C^{1,1}$. Moreover, η, κ are physical parameters, and $\mathbf{H}$ is a fixed externally applied magnetic field that is dependent on time. The *supercurrent density* $\mathbf{J}_s$ is given by

$$\mathbf{J}_s := -\mathrm{Re}\left[\psi^*\left(\frac{i}{\kappa}\nabla + \mathbf{A}\right)\psi\right]. \tag{4.67b}$$

(See Tinkham [**119**] for the terminology.) The associated boundary conditions are

$$\begin{cases} \mathbf{n}\cdot\nabla\psi = 0 = \mathbf{n}\cdot\mathbf{A}, \\ \mathbf{n}\times(\nabla\times\mathbf{A} - \mathbf{H}) = 0 \end{cases} \quad \text{on} \quad \partial\Omega. \tag{4.67c}$$

The corresponding Ginzburg-Landau (GL) free energy

$$\begin{aligned} \mathcal{E}_{\mathbf{H}}(\psi,\mathbf{A}) := \frac{1}{2}\int_\Omega \left[\left|\left(\frac{i}{\kappa}\nabla + \mathbf{A}\right)\psi\right|^2 + |\nabla\times\mathbf{A} - \mathbf{H}|^2\right] dx \\ +\mathcal{F}(\psi) \quad \text{with} \quad \mathcal{F}(\psi) := \frac{1}{2}\int_\Omega (|\psi|^2 - 1)^2\, dx \end{aligned} \tag{4.68a}$$

is well-defined on the Banach space

$$\mathcal{X} := [W^{1,2}(\Omega)]^2 \times [W^{1,2}(\Omega)]^N_{\mathbf{n},0}, \tag{4.68b}$$

where $[W^{1,2}(\Omega)]^N_{\mathbf{n},0}$ denotes the closed linear subspace of the Hilbert space $[W^{1,2}(\Omega)]^N$, which consists of all vector-valued functions $\mathbf{A} : \Omega \to \mathbb{R}^N$ from $[W^{1,2}(\Omega)]^N$ such that $\mathbf{n} \cdot \mathbf{A} = 0$ on the boundary in the sense of trace.

Set $\eta = 1$. Then the (TDGL) equations (4.67a,b,c) can be rewritten as the *time-dependent gradient system* generated by the GL energy $\mathcal{E}_{\mathbf{H}}$:

$$\dot{v} = -\mathcal{E}'_{\mathbf{H}}(v),\ t \geq 0; \quad v = v(t) \in \mathcal{X}. \tag{4.68c}$$

We impose the first condition on the applied magnetic field $\mathbf{H}$:

(h1) There exists a constant $\alpha \in (1/2, 1)$ such that

$$\mathbf{H} \in L^\infty(0,T;[W^{\alpha,2}(\Omega)]^N) \cap W^{1,2}(0,T;[L^2(\Omega)]^N) \quad \forall T > 0.$$

This condition (**h1**) ensures the global existence of solutions v of (4.67) for sufficiently smooth initial data $v(0) = v_0$. For instance, it is known [**42**, Theorem 3 (ii)] that for every initial value $v_0 \in \mathcal{D}(\mathcal{A}^{(1+\alpha)/2})$, the equation (4.67) has a unique solution v whose orbit has compact closure in

$$\mathcal{W}^{1+\alpha,2} := [W^{1+\alpha,2}(\Omega)]^2 \times [W^{1+\alpha,2}(\Omega)]^N,$$

where $\mathcal{A}$ is a linear self-adjoint operator in the Hilbert space

$$\mathcal{L}^2 := [L^2(\Omega)]^2 \times [L^2(\Omega)]^N$$

associated with some quadratic form Q; the exact form of Q can be found in Fleckinger-Kaper-Takáč [**42**] or Feireisl-Takáč [**41**]. Note that the space $\mathcal{X}$ is compactly embedded in $\mathcal{W}^{1+\alpha,2}$. Therefore, the orbit of such a solution also has a compact closure in $\mathcal{X}$. In other words, the assertion in [**42**, Theorem 3 (ii)] implies that solutions of (4.67) with initial data $v_0 \in \mathcal{D}(\mathcal{A}^{(1+\alpha)/2})$ have precompact orbits in the space $\mathcal{V} = \mathcal{X}$. Thus, by virtue of Theorem 4.1 in Chapter 3, there holds the convergence in $\mathcal{X}$ of such a solution if we can rewrite the term $\mathcal{E}'_{\mathbf{H}}(v)$ on the r.h.s. of (4.67a) as the sum $\mathcal{E}'_\infty(v) + f(t,v)$, where $\mathcal{E}_\infty$ is a time-independent analytic energy functional and the perturbation $f(t,v)$ is sufficiently small.

For this purpose, we impose the following additional condition on the time-dependent applied magnetic field $\mathbf{H}$:

(h2): There exists a time-independent applied magnetic field $\mathbf{H}_\infty \in [W^{1,2}(\Omega)]^N$ such that

$$\mathbf{n} \times (\mathbf{H} - \mathbf{H}_\infty) = 0 \quad \text{a.e. on} \quad \partial\Omega \in (0,\infty), \tag{4.69a}$$

and there exists some increasing positive function ρ with $\frac{1}{\rho} \in L^1$ such that

$$\int_0^\infty \rho(t) \|\nabla \times (\mathbf{H}(t) - \mathbf{H}_\infty)\|^2_{L^2}\, dt < \infty. \tag{4.69b}$$

We denote by $\mathcal{E}_\infty$ the GL energy functional $\mathcal{E}_{\mathbf{H}_\infty}$ corresponding to the time-independent magnetic field $\mathbf{H}_\infty$. It is shown [**41**, Proposition 6.1] that $\mathcal{E}_\infty$ is analytic on $\mathcal{X}$ and near each critical point its gradient $\mathcal{E}'_\infty$ satisfies a gradient inequality of LS-type, with respect to the dual $\mathcal{X}'$ norm and thus to the $\mathcal{L}^2$ norm.

We define the map $g : \mathbb{R}_+ \times \mathcal{X} \to \mathcal{L}^2$ by

$$g(t,v) := \mathcal{E}'_\infty(v) - \mathcal{E}'_{\mathbf{H}}(v) \quad \forall t \geq 0; v = (\psi, \mathbf{A}) \in \mathcal{X}.$$

Note that

$$\mathcal{E}_\infty(v) - \mathcal{E}_{\mathbf{H}}(v) = \frac{1}{2}\int_\Omega \left[|\nabla \times \mathbf{A} - \mathbf{H}_\infty|^2 - |\nabla \times \mathbf{A} - \mathbf{H}(t)|^2\right] dx$$

$$= \int_\Omega (\nabla \times \mathbf{A}) \cdot (\mathbf{H}(t) - \mathbf{H}_\infty)\, dx + \frac{1}{2}\int_\Omega \left(|\mathbf{H}_\infty|^2 - |\mathbf{H}(t)|^2\right) dx.$$

The second sum term is independent of v. To calculate the first sum term, we use the elementary identity

$$(\nabla \times \mathbf{F}) \cdot \mathbf{G} = \mathbf{F} \cdot (\nabla \times \mathbf{G}) + \operatorname{div}(\mathbf{F} \times \mathbf{G}).$$

It yields that the first sum term is equal to

$$\int_\Omega \mathbf{A} \cdot (\nabla \times (\mathbf{H}(t) - \mathbf{H}_\infty))\, dx + \int_\Omega \operatorname{div}(\mathbf{A} \times (\mathbf{H}(t) - \mathbf{H}_\infty))\, dx.$$

We use the classical divergence theorem to find that

$$\int_\Omega \operatorname{div}(\mathbf{A} \times (\mathbf{H}(t) - \mathbf{H}_\infty))\, dx = \int_{\partial\Omega} \mathbf{n} \cdot (\mathbf{A} \times (\mathbf{H}(t) - \mathbf{H}_\infty))\, dF$$

$$= \int_{\partial\Omega} \mathbf{A} \cdot (\mathbf{n} \times (\mathbf{H}(t) - \mathbf{H}_\infty))\, dF.$$

The latter integrand is zero by condition (4.69a). Therefore,

$$g(t,v) = (0, \nabla \times (\mathbf{H}(t) - \mathbf{H}_\infty)) =: G(t)$$

is independent of v and

$$\|G(t)\|_{\mathcal{L}^2} \leq \|\nabla \times (\mathbf{H}(t) - \mathbf{H}_\infty)\|_{L^2}, \quad t \geq 0. \tag{4.70}$$

Under the conditions **(h1)-(h2)** we let v be a solution of (4.67) such that its orbit $O(v) := \{v(t) : t \geq 0\}$ is precompact in $\mathcal{W}^{1+\alpha,2}$ and thus also precompact in $\mathcal{X}$. Then v solves the following perturbed gradient system

$$\dot{v} = -\mathcal{E}'_\infty(v) + G(t), \quad t \geq 0 \tag{4.71}$$

in the Hilbert space $\mathcal{L}^2$. Using (4.70) combined with the condition (4.69b) we see that (4.71) is a perturbed gradient system with which we have dealt in §4 of Chapter 3, where the corresponding triplet is

$$(H, X, \mathcal{E}') = (\mathcal{L}^2, \mathcal{X}, \mathcal{E}'_\infty),$$

and the corresponding condition (3.28c′) on the perturbation G is merely the given condition (4.69b) by noting (4.70). Moreover, as mentioned before, $\mathcal{E}'_\infty$ satisfies a gradient inequality of LS-type near each cluster point of $O(v)$ in $\mathcal{X}$, with respect to the $\mathcal{X}'$ norm. Therefore, for the compact solution v, our Theorem 4.1 (ii) of Chapter 3 applies and yields the convergence $\|v(t) - \varphi\|_{\mathcal{X}} \to 0$ as $t \to \infty$ for some critical point $\varphi \in \mathcal{X}$. This implies, by the precompactness of the orbit $O(v)$ in $\mathcal{W}^{1+\alpha,2}$, that $\varphi \in \mathcal{W}^{1+\alpha,2}$ and the convergence $\|v(t) - \varphi\|_{\mathcal{W}^{1+\alpha,2}} \to 0$ as $t \to \infty$.

The above convergence result applies particularly to the case where the externally applied magnetic field $\mathbf{H}$ satisfies condition (4.69b) in (**h2**) in the sense that

$$\int_0^\infty t|\log t|^\tau \|\nabla \times (\mathbf{H}(t) - \mathbf{H}_\infty)\|_{L^2}^2\, dt < \infty \quad \text{with some} \quad \tau > 1. \tag{4.69b$'$}$$

Note that (4.69b′) does not imply, in general, the convergence $\|\nabla \times (\mathbf{H}(t) - \mathbf{H}_\infty)\|_{L^2}$ as $t \to \infty$.

5. Convergence in porous medium models

In this section we will sketch the proof of a convergence result for the porous medium equations given by E. Feireisl and F. Simondon [**40**].

EXAMPLE 5.1. **(Porous medium equations)** In [**40**], E. Feireisl and F. Simondon establish the convergence result for positive solutions of the porous medium equations

$$\begin{cases} \dot{u} = \Delta u^m - f(u) \quad \text{on} \quad \Omega, \\ u = 0 \quad \text{on} \quad \partial\Omega, \end{cases} \tag{4.72a}$$

where $m > 0$ and the function $f \in C([0, \infty))$ has the following property: $f(0) \leq 0$ and there exists a real analytic function $\tilde{f}$ on $\mathbb{R}$ such that

$$f(u) = \tilde{f}(u^m) \quad \forall u \geq 0. \tag{4.72b}$$

We remark that the condition on f that we have imposed as above is stronger than the ones used by Feireisl and Simondon [**40**].

The convergence result of Feireisl and Simondon [**40**, Theorem 3.1] reads: If u is a weak solution of (4.72) such that

$$0 \leq u(t, \cdot) \leq M \quad \text{for} \quad t > 0,$$

where M is a constant, then u is continuous for all $t > 0$ and there exists a classical stationary solution w of (4.72) such that

$$\lim_{t\to\infty} \|u(t) - w\|_{C(\bar{\Omega})} = 0.$$

Below we will sketch how they deduce this convergence result from certain gradient inequalities.

As observed in [**40**, Prop. 2.2], the above positivity and boundedness of the solution u implies that for each $t_0 > 0$, the section $\{u(t) : t \geq t_0\}$ is precompact in $C(\bar{\Omega})$. Hence, there are only two possibilities: (a) The precompact section $\{u(t) : t \geq t_0\}$ in $C(\bar{\Omega})$ has a cluster point w which is nonzero, or (b) $u(t) \to 0$ as $t \to \infty$ in $C(\bar{\Omega})$.

We need to treat case (a). In this case, condition (4.72c) implies [**40**, Prop. 4.3] that $u(t, x) > 0$ for all $t \geq t_0$ and $x \in \Omega$. Hence, without loss of generality, we can assume that the orbit $O(u) = \{u(t) : t \geq 0\}$ itself is a precompact subset of $C(\bar{\Omega})$ and $u(t, x) > 0$ for all $t \geq 0, x \in \Omega$. By setting $v := u^m$, we see that v solves the following equations:

$$\dot{u} = \Delta v - \tilde{f}(v), \quad v = u^m \geq 0. \tag{4.73}$$

Here $\tilde{f}$ is the real analytic function such that (4.72b) holds.

Set $V := W_0^{1,2}(\Omega)$ and consider on $V \cap L^\infty$ the functional

$$\mathcal{E}_{\tilde{f}}(v) := \frac{1}{2}\int_\Omega |\nabla v|^2\,dx + \int_\Omega F(v)\,dx \quad (F' = \tilde{f}).$$

The analyticity of $\tilde{f}$ ensures the applicability of Theorem 3.1 to the gradient map $\mathcal{E}'_{\tilde{f}}$ given by $\mathcal{E}'_{\tilde{f}}(v) := -\Delta v + \tilde{f}(v)$ ($\forall v \in V \cap L^\infty$).

Let $\varphi := w^m$, where w is a nonzero cluster point of the precompact orbit $\{u(t) : t \geq 0\}$ in $C(\bar{\Omega})$. φ is a cluster point of the precompact orbit $O(v) = \{v(t) : t \geq 0\}$ in $C(\bar{\Omega})$ and thus is a critical point of $\mathcal{E}_{\tilde{f}}$. It follows that $\varphi \in W_0^{1,2}(\Omega) \cap C(\bar{\Omega}) \subset V \cap L^\infty$ by virtue of the classical regularity result. Furthermore, the classical Maximum Principle yields that

$$\varphi(x) \geq \text{const} \cdot \text{dist}(x, \partial\Omega) \quad \forall x \in \Omega. \tag{4.74a}$$

As mentioned before, Theorem 3.1 (i) applies to the gradient map $\mathcal{E}'_{\tilde{f}}$. It yields for the critical point φ that there are constants $c, \sigma > 0$ and $\theta \in [1/2, 1)$ such that

$$\begin{cases} c \cdot |\mathcal{E}_{\tilde{f}}(v(t)) - \mathcal{E}_{\tilde{f}}(\varphi)|^\theta \leq \|\dot{u}(t)\|_{W^{-1,2}} & \forall t \in \Lambda, \\ \Lambda := \{t \geq 0 : \|v(t) - \varphi\|_{L^\infty} < \sigma\}. \end{cases} \tag{4.74b}$$

In the above, we have replaced $\mathcal{E}'_{\tilde{f}}(v(t))$ with $-\dot{u}(t)$.

To continue, we use the following energy inequality for the solution u (see [**40**]):

$$\int_{t_1}^{t_2} \|\dot{w}(t)\|_{L^2}^2\,dt \leq \mathcal{E}_{\tilde{f}}(v(t_1)) - \mathcal{E}_{\tilde{f}}(v(t_2)), \quad 0 \leq t_1 < t_2, \tag{4.75a}$$

where

$$w(t) := \frac{2\sqrt{m}}{m+1} u(t)^{\frac{1+m}{2}} \quad (t \geq 0). \tag{4.75b}$$

It follows from (4.75a) that

$$\lim_{t\to\infty} \mathcal{E}_{\tilde{f}}(v(t)) = \mathcal{E}_{\tilde{f}}(\varphi)$$

and there holds

$$\int_t^\infty \|\dot{w}(s)\|_{L^2}^2\,ds \leq \mathcal{E}_{\tilde{f}}(v(t)) - \mathcal{E}_{\tilde{f}}(\varphi) \quad \forall t \geq 0. \tag{4.75c}$$

This implies by (4.74b) that

$$c \cdot \left(\int_t^\infty \|\dot{w}(s)\|_{L^2}^2\,ds\right)^\theta \leq \|\dot{u}(t)\|_{W^{-1,2}} \quad \forall t \in \Lambda. \tag{4.76a}$$

The final step is to show that there exists some constant $c_1 > 0$ such that

$$\|\dot{w}(s)\|_{L^2} \geq c_1 \cdot \|\dot{u}(s)\|_{W^{-1,2}} \quad \forall s \in \Lambda. \tag{4.76b}$$

Note that this is equivalent to the existence of some positive constant c_1 such that

$$\|v(s)^{\frac{m-1}{2m}}\dot{u}(s)\|_{L^2} \geq c_1 \cdot \|\dot{u}(s)\|_{W^{-1,2}} \quad \forall s \in \Lambda. \tag{4.76b$'$}$$

For the case $m \leq 1$, (4.76b$'$) follows directly from the boundedness of the solution u. For the case $m > 1$, the proof of (4.76b$'$) given in [**40**, Lemma 7.3] is more involved; in the initial step one needs to prove that all of the points $v(t) = u(t)^m$ with $t \in \Lambda$ satisfy an estimate of the form (4.74a) with a common constant, once the positive constant σ has been taken to be sufficiently small. In the final step one needs to use the Hardy inequality to derive the desired estimates in (4.76b$'$).

Combining (4.76a) with (4.76b), we find that the L^2 function h given by $h(t) := \|\dot{u}(t)\|_{W^{-1,2}}$ for $t \geq 0$ is such that

$$\phi\left(\int_t^\infty h(s)^2\, ds\right) \leq h(t) \quad \forall t \in \Lambda,$$

where the function

$$\phi(s) := \min\{cc_1^\theta, 1\} \cdot |s|^\theta \quad (s \in \mathbb{R})$$

is in the class $\mathcal{G}$. Therefore, by Lemma 2.2 of Chapter 3 under the choice $g \equiv 0$ and $f = h$, we conclude that $\int_\Lambda h(t)\, dt < \infty$, i.e.,

$$\int_\Lambda \|\dot{u}(t)\|_{W^{-1,2}}\, dt = \int_\Lambda h(t)\, dt < \infty.$$

Therefore, since the section $\{u(t) : t \geq t_0\}$ is precompact in $C(\bar{\Omega})$, an application of Lemma 2.3 of Chapter 3 yields the desired convergence $\|u(t) - \varphi\|_{C(\bar{\Omega})} \to 0$ as $t \to \infty$.

In the above result, the dimension of the bounded domain Ω is arbitrary. For the one dimensional case $\Omega = (-L, L)$, Feireisl and Simondon [**39**] have also established the convergence to one single equilibrium for positive and bounded solutions of (4.72) under the following even weaker condition on the function f : $f \in C^1((0, \infty))$ and $f(0) \leq 0$.

CHAPTER 5

Applications to the stability problem

In previous chapters we have seen that the validity of gradient inequalities ensures stability of the ground states for gradient flows on Hilbert spaces generated by energy functionals. A gradient flow gives the *steepest descent* of the functional along its flow-lines. However, in many application situations, it is not at all necessary to deform along lines of "steepest" descent in order to find the extremum of the given functional; "steep enough" suffices and the widely used tool is the so-called pseudo-gradient vector fields [**112**].

The advantage of using pseudo-gradient vector fields is twofold: the first one is the flexibility and the second one is that the corresponding evolution equations that one must solve are ordinary differential equations on Banach spaces for which there is a complete theory. Certainly, the latter point is more important for practical applications.

Our study in this chapter focuses on the convergence of the flows generated by pseudo-gradient vector fields near a given ground state, to which the following stability problems are tackled: the usual Lyapunov stability of the given ground state with respect to the dynamical system generated by a pseudo-gradient vector field and the stability of the convergence property of the solutions to these "pseudo-gradient" dynamical systems under "small" circumstance perturbations.

As it will be seen below, gradient inequalities play a crucial rôle for solving the above-mentioned stability problems.

1. Stability of ground states

Let us consider the following time-dependent evolution equation

$$\dot{u}(t) = -\mathcal{N}(t, u(t)), t \geq 0; \quad u(0) = u_0 \in X \tag{5.1a}$$

settled in a real Banach space X. Here, we assume that $\mathcal{N} : \mathbb{R}_+ \times X \to X$ is a continuous map that obeys the so-called "local Lipschitz property": For each $T > 0$ and $C > 0$ there exists a positive constant $L \equiv L(T, C)$ such that

$$\|\mathcal{N}(t, u) - \mathcal{N}(t, v)\|_X \leq L \cdot \|u - v\|_X \tag{5.1b}$$

for all $t \in [0, T]$ and $u, v \in X$ with $\|u\|_X \leq C$, $\|v\|_X \leq C$. It follows from the basic differential equation theory that this local Lipschitz property ensures the following local existence: For each $u_0 \in X$, the corresponding Cauchy problem (5.1a) has a classical solution that exists in a time interval $[0, T)$.

Our idea of establishing the long-time convergence of global solutions of (5.1a) is to compare the system (5.1a) with the pseudo-gradient system generated by some C^1 functional that satisfies a certain gradient inequality. The tools that we will use are the results established in §3 of Chapter 3.

We let $U \subset X$ be an open subset of X and consider a C^1 functional $\mathcal{E} : U \to \mathbb{R}$ so that the Fréchet derivative $\mathcal{E}'$ is a continuous map from U into X', the dual of X. Let $\mathcal{S} := \{\varphi \in U : \mathcal{E}'(\varphi) = 0\}$ be the set of critical points of $\mathcal{E}$. We consider a critical point $\varphi \in \mathcal{S}$ that is a *ground state* of $\mathcal{E}$ in the sense that $\mathcal{E}$ attains its minimum at this point, i.e.,

$$\mathcal{E}(\varphi) = \inf_{u \in U} \mathcal{E}(u). \tag{5.2}$$

We take $Y = X'$ so that $\mathcal{M} = \mathcal{E}' : U \to Y = X'$ is a gradient map. It follows that the semi-norm $\|\cdot\|_{Y'}$ on X is just the usual norm on X :

$$\|u\|_{Y'} = \sup\{\langle u, y\rangle_{X \times X'} : y \in Y = X', \|y\|_{X'} \leq 1\} = \|u\|_X.$$

In the following we use U_r to denote an r-neighborhood of φ, i.e.,

$$U_r := \{u \in X : \|u - \varphi\|_X < r\}.$$

In order to use the gradient field $\mathcal{E}'$ to study the system (5.1a), we fix a constant $\sigma > 0$ such that $U_\sigma \subset U$ and impose on the map $\mathcal{N}$ a local growth condition of the form:

$$\begin{cases} \langle \mathcal{E}'(u), \mathcal{N}(t,u)\rangle_{X' \times X} \geq \alpha \cdot \|\mathcal{E}'(u)\|_{X'} \cdot \|\mathcal{N}(t,u)\|_X + F'(t); \\ \|\mathcal{E}'(u)\|_{X'} \geq \alpha \cdot \|\mathcal{N}(t,u)\|_X + G'(t) \quad \forall (t,u) \in \mathbb{R}_+ \times U_\sigma, \end{cases} \tag{5.3}$$

where $\alpha > 0$ is a positive constant and $F, G : \mathbb{R}_+ \to \mathbb{R}_+$ are nonincreasing C^1 functions whose smallness will be determined later (see (5.5) below). Our first convergence result reads as follows.

THEOREM 1.1. *Let φ be a ground state of $\mathcal{E}$ and assume $\mathcal{E}(\varphi) = 0$. Assume the following gradient inequality of type $\mathcal{G}_k$:*

$$\phi(\mathcal{E}(v)) \leq \|\mathcal{E}'(v)\|_{X'} \quad \forall v \in U_\sigma, \tag{5.4a}$$

where σ is a positive constant such that $U_\sigma \subset U$ and $\phi \in \mathcal{G}_k$ with some $k \geq 1$, i.e., $0 \leq \frac{1}{\phi} \in L^1_{loc}(\mathbb{R})$ and there holds

$$\phi(x+y) \leq k\phi(x) + |y|^{1/2} \quad \forall x, y \in \mathbb{R}. \tag{5.4b}$$

Consider a system (5.1a) where the map $\mathcal{N}$ satisfies the growth condition (5.3) with two nonincreasing C^1 functions F and G that are so small that

$$\begin{cases} \lim_{t\to\infty} (F(t) + G(t)) = 0 \quad \text{and} \\ \delta_1 := G(0) + \displaystyle\int_0^{F(0)} \frac{ds}{\phi(s)} + \int_0^\infty \sqrt{F(t)}\,dt < \frac{\alpha\sigma}{2k}. \end{cases} \tag{5.5}$$

Let $\varepsilon \in (0, \sigma/2)$ be so small that

$$G(0) + \int_0^{\mathcal{E}(u)+F(0)} \frac{ds}{\phi(s)} + \int_0^\infty \sqrt{F(t)}\,dt < \frac{\alpha\sigma}{2k} \quad \forall u \in U_\varepsilon. \tag{5.6}$$

Then, for each $u_0 \in U_\varepsilon$*, the Cauchy problem (5.1a) admits a global solution* $u : [0,\infty) \to U_{\sigma/2}$ *that converges as* $t \to \infty$ *with respect to the* X *norm in the sense that* $I(u) = \int_0^\infty \|\dot{u}(t)\|_X < \infty$.

PROOF. We note that the existence of such a positive constant ε with the described property is guaranteed by the condition $\delta_1 < \alpha\sigma/(2k)$ in (5.5) combined with the assumption $\mathcal{E}(\varphi) = 0$.

Fix $u_0 \in U_\varepsilon$. Then the corresponding Cauchy problem (5.1a) admits a solution u that exists on a maximal interval $[0, \tilde{T})$. We want to show that

$$\|u(t) - \varphi\|_X < \sigma/2 \quad \forall t \in [0, \tilde{T}). \tag{5.7}$$

Note that (5.7) implies $\tilde{T} = \infty$ by the maximal property of $\tilde{T}$ and thus the solution u exists globally on $[0,\infty)$. More importantly, (5.7) allows us to apply the gradient inequality (5.4a), and thus an application of Theorem 3.6 of Chapter 3 will yield the desired convergence of the solution u.

We give an argument by contradiction. Suppose there exists $T \in (0, \tilde{T})$ such that

$$\|u(t) - \varphi\|_X < \sigma/2 \quad \forall t \in [0, T) \qquad \text{but} \qquad \|u(T) - \varphi\|_X \geq \sigma/2. \tag{5.8}$$

Using $\mathcal{N}(t, u(t)) = -\dot{u}(t)$, we rewrite condition (5.3) as

$$\begin{cases} \langle -\mathcal{E}'(u(t)), \dot{u}(t)\rangle_{X' \times X} \geq \alpha \cdot \|\mathcal{E}'(u(t))\|_{X'} \cdot \|\dot{u}(t)\|_X + F'(t); \\ \|\mathcal{E}'(u(t))\|_{X'} \geq \alpha \cdot \|\dot{u}(t)\|_X + G'(t) \quad \forall t \geq 0. \end{cases} \tag{5.9}$$

This implies that u is a trajectory satisfying a growth condition of the form (3.10a) in §3 of Chapter 3, under the choice of the space $Y = X'$ and the linear operator $\gamma = \alpha \cdot \text{Identity}$. Moreover, the gradient inequality (5.4a) applies to each $v = u(t) \in U_\sigma$ with $t \leq T$ and yields that

$$\phi(\mathcal{E}(u(t))) \leq \|\mathcal{E}'(u(t))\|_{X'} \quad \forall t \in [0, T].$$

Therefore, we are in the position to apply Lemma 3.4 of Chapter 3 and find that (see (3.18b))

$$(\alpha/k) \cdot \int_0^T \|\dot{u}(t)\|_X \, dt \leq G(0) + \int_0^T \sqrt{F(t)} \, dt + \int_{H(T)}^{H(0)} \frac{ds}{\phi(s)}. \tag{5.10}$$

Here, $H(t) := \mathcal{E}(u(t)) + F(t)$ $(t \geq 0)$. We have $F(T) \geq 0$ by assumption (5.5) combined with the monotonicity of F. Moreover, $\mathcal{E}(u(T)) \geq \mathcal{E}(\varphi) = 0$, since $u(T) \in U_\sigma \subset U$ and φ is a ground state. Hence, $H(T) = \mathcal{E}(u(T)) + F(T) \geq 0$, $H(0) = \mathcal{E}(u_0) + F(0)$ and thus the term on the r.h.s. of (5.10) is bounded by the constant $\alpha\sigma/(2k)$ (see (5.6)). Consequently, $\int_0^T \|\dot{u}(t)\|_X \, dt < \sigma/2$. It follows that

$$\|u_0 - u(T)\|_X \leq \int_0^T \|\dot{u}(t)\|_X \, dt < \sigma/2$$

and thus

$$\varepsilon > \|u_0 - \varphi\|_X \geq \|u(T) - \varphi\|_X - \|u_0 - u(T)\|_X > \sigma - \sigma/2 = \sigma/2,$$

contradicting our choice of $\varepsilon \in (0, \sigma/2)$.

The above shows that $\tilde{T} = \infty$ and (5.7) holds true with $\tilde{T} = \infty$. Hence, we have proved that the solution $u(t)$ exists for all $t \geq 0$ and its orbit $O(u)$ is contained in the set U_σ so that the gradient inequality (5.4a) applies to every point $v = u(t)$. Therefore, Theorem 3.6 of Chapter 3 applies and yields the convergence of the integral $I(u) < \infty$, which implies the convergence of u in the norm of X. This finishes the proof. □

THEOREM 1.2. **(Stability Theorem)** *Let φ be a ground state of $\mathcal{E}$ with $\mathcal{E}(\varphi) = 0$ and assume the following gradient inequality:*

$$\phi(\mathcal{E}(u)) \leq \|\mathcal{E}'(u)\|_{X'} \quad \forall u \in U_\sigma \tag{5.11}$$

for some positive constant σ such that $U_\sigma \subset U$ and some function ϕ with $0 \leq \frac{1}{\phi} \in L^1_{loc}(\mathbb{R})$.

Let $\mathcal{N} : U \to X$ be a local Lipschitz map for which there exists a constant $\alpha > 0$ such that

$$\begin{cases} \langle \mathcal{E}'(u), \mathcal{N}(u)\rangle_{X' \times X} \geq \alpha \cdot \|\mathcal{E}'(u)\|_{X'} \cdot \|\mathcal{N}(u)\|_X, \\ \|\mathcal{E}'(u)\|_{X'} \geq \alpha \cdot \|\mathcal{N}(u)\|_X \end{cases} \tag{5.12}$$

for all $u \in U$. Choose $\varepsilon \in (0, \sigma/2)$ to be so small that

$$\int_0^{\mathcal{E}(u)} \frac{ds}{\phi(s)} < (\alpha\sigma)/2 \quad \forall u \in U_\varepsilon. \tag{5.13}$$

Then, for each $u_0 \in U_\varepsilon$, the following autonomous Cauchy problem

$$\begin{cases} \dot{u}(t) = -\mathcal{N}(u(t)), t \geq 0; \\ u(0) = u_0 \end{cases} \tag{5.14}$$

admits a global solution $u : [0, \infty) \to U_{\sigma/2}$ that converges as $t \to \infty$ to some stationary point $u_\infty \in \mathcal{N}^{-1}(0) \cap U_\sigma$ with respect to the X norm. As a consequence, φ as an equilibrium of (5.14) is Lyapunov stable.

Moreover, φ is uniformly asymptotically stable if, in addition, the following conditions (i)–(ii) are fulfilled:

(i) There holds

$$\inf\{\|\mathcal{E}'(u)\|_{X'} : u \in U_\sigma, \mathcal{E}(u) = r\} > 0 \tag{5.15a}$$

for all sufficiently small but positive r.

(ii) The inequality in the second line of (5.12) is strengthened by

$$(1/\alpha) \cdot \|\mathcal{N}(u)\|_X \geq \|\mathcal{E}'(u)\|_{X'} \geq \alpha \cdot \|\mathcal{N}(u)\|_X \quad (\forall u \in U). \tag{5.15b}$$

PROOF. The existence of global solutions $u : [0, \infty) \to U_{\sigma/2}$ to (5.14) can be proved in the same spirit as that of Theorem 1.1. Here we omit the details. Now for any such global solution u we have $u(t) \in U_\sigma$ and thus, by the gradient inequality (5.11),

$$\phi(\mathcal{E}(u(t))) \leq \|\mathcal{E}'(u(t))\|_{X'} \quad \forall t \geq 0.$$

From this we see that u is a trajectory satisfying (3.10′) in §3 of Chapter 3 under the choice $Y = X'$ and $\gamma = \alpha \cdot \text{Identity}$. Therefore, Proposition 3.2 of Chapter 3 applies and yields (see (3.15b)) the estimate

$$\alpha \cdot \int_t^\infty \|\dot{u}(s)\|_X \, ds \le \int_a^{\mathcal{E}(u(t))} \frac{d\tau}{\phi(\tau)} \quad \forall t \ge 0, \tag{5.16a}$$

where $a := \lim_{t\to\infty} \mathcal{E}(u(t))$. Note that (5.16a) implies the convergence of the solutions $u(t)$ as $t \to \infty$ to some equilibrium $u_\infty \in U_\sigma$.

To prove the Lyapunov stability of φ, we repeat the above argument for each fixed $\tau \in (0, \sigma)$. It yields that for each $\tau \in (0, \sigma)$ there exists $\varepsilon(\tau) \in (0, \tau/2)$ such that for each $u_0 \in U_{\varepsilon(\tau)}$, Eq. (5.14) has a global solution u such that

$$\|u(t) - \varphi\|_X < \tau/2 \quad \forall t \ge 0.$$

This shows the desired Lyapunov stability of φ.

It remains to prove the uniform asymptotic stability of φ given in the theorem. For this purpose, we may assume that $\sigma > 0$ has been chosen to be so small that

$$\beta := \sup\{\mathcal{E}(u) : u \in U_\sigma\} < \infty$$

and that (5.15a) holds true for any $r \in (0, \beta]$. Hence, the validity of (5.15a) for all $r \in (0, \beta]$ implies that $\mathcal{E}$ has no other equilibria of $\mathcal{E}$ in U_σ than φ. This yields by (5.15b) that φ is the unique equilibrium of (5.14) in U_σ.

We define a function $\tilde{\phi}$ on $[0, \beta)$ by setting $\tilde{\phi}(0) = 0$ and

$$\tilde{\phi}(r) := \inf\{\|\mathcal{E}'(u)\|_{X'} : u \in U_\sigma, \mathcal{E}(u) = r\} > 0$$

for all $r \in (0, \beta]$. We have that

$$\tilde{\phi}(\mathcal{E}(v)) \le \|\mathcal{E}'(v)\|_{X'} \quad \forall v \in U_\sigma \tag{$*$}$$

and $\tilde{\phi}(r) \ge \phi(r)$ for all $r \in [0, \beta]$. Since $1/\phi \in L^1_{loc}(\mathbb{R})$, it follows that $1/\tilde{\phi} \in L^1([0, \beta])$ and thus we can extend $\tilde{\phi}$ to be a function on $\mathbb{R}$ such that $0 \le 1/\tilde{\phi} \in L^1_{loc}(\mathbb{R})$.

By virtue of $(*)$ we find that the function ϕ involved in (5.16a) can be replaced by the bigger function $\tilde{\phi}$; i.e., we have

$$\alpha \cdot \int_t^\infty \|\dot{u}(s)\|_X \, ds \le \int_{\mathcal{E}(u_\infty)}^{\mathcal{E}(u(t))} \frac{1}{\tilde{\phi}(s)} \, ds \quad \forall t \ge 0, \tag{5.16a′}$$

where $u_\infty := \lim_{t\to\infty} u(t)$ and the limit is taken in X. Clearly, $\tilde{\phi}$ is lower semicontinuous. Hence, $\tilde{\phi}$ is strictly positive on any compact subset of $(0, \beta]$ and thus for each $t \ge 0$ there exists a unique number $g(t) \in (0, \beta]$ such that

$$\alpha^2 t = \int_{g(t)}^{\beta} \frac{ds}{\tilde{\phi}(s)^2}. \tag{5.16b}$$

Moreover, there holds

$$g(t) \to 0 \quad \text{as} \quad t \to \infty. \tag{5.16c}$$

To continue, we consider a global solution $u : [0, \infty) \to U_\sigma$ of (5.14) and put $h(t) := \mathcal{E}(u(t))$ for $t \geq 0$. Then

$$\begin{aligned} -h'(t) &= \langle \mathcal{E}'(u(t)), -\dot{u}(t)\rangle_{X' \times X} = \langle \mathcal{E}'(u(t)), \mathcal{N}(u(t))\rangle_{X' \times X} \\ &\geq (\alpha \|\mathcal{E}'(u(t))\|_{X'}) \cdot \|\mathcal{N}(u(t))\|_X \quad \text{(by (5.12))} \\ &\geq (\alpha \|\mathcal{E}'(u(t))\|_{X'}) \cdot (\alpha \|\mathcal{E}'(u(t))\|_{X'}) \quad \text{(by (5.15b))} \end{aligned}$$

and thus

$$-h'(t) \geq \alpha^2 \cdot \|\mathcal{E}'(u(t))\|_{X'}^2. \tag{5.16d}$$

We apply the above gradient inequality $(*)$ to each point $v = u(t) \in U_\sigma$ and find that

$$\|\mathcal{E}'(u(t))\|_{X'} \geq \tilde{\phi}(\mathcal{E}(u(t))) = \tilde{\phi}(h(t)).$$

It follows from (5.16d) that

$$\frac{-h'(t)}{\tilde{\phi}(h(t))^2} \geq \alpha^2 \quad \forall t \geq 0. \tag{5.16e}$$

By integrating (5.16e) over any finite interval $[0, t]$, we find that

$$\alpha^2 t \leq \int_{h(t)}^{h(0)} \frac{ds}{\tilde{\phi}(s)^2} \quad \forall t \geq 0. \tag{5.16f}$$

Since $h(0) = \mathcal{E}(u_0) \leq \beta$, we obtain by comparing (5.16f) with (5.16b) that

$$\mathcal{E}(u(t)) = h(t) \leq g(t) \quad \forall t \geq 0.$$

This implies by (5.16a′) that

$$\|u(t) - u_\infty\|_X \leq \int_t^\infty \|\dot{u}(\tau)\|_X \, d\tau \leq (1/\alpha) \cdot \int_0^{g(t)} \frac{ds}{\tilde{\phi}(s)} \quad \forall t \geq 0$$

with $u_\infty := \lim_{t \to \infty} u(t)$ an equilibrium of (5.14) in U_σ and $\mathcal{E}(u_\infty) \geq \mathcal{E}(\varphi) = 0$. But φ is the unique equilibrium of (5.14) in the neighborhood U_σ. Therefore, $u_\infty = \varphi$ and thus

$$\|u(t) - \varphi\|_X \leq (1/\alpha) \cdot \int_0^{g(t)} \frac{1}{\tilde{\phi}(\tau)} \, d\tau =: \gamma(t) \quad (\forall t \geq 0). \tag{5.16g}$$

The convergence $\gamma(t) \to 0$ as $t \to \infty$ follows from the local integrability of $1/\tilde{\phi}$ combined with the convergence $g(t) \to 0$ as $t \to \infty$ in (5.16c). The uniform asymptotic stability of φ follows directly from the estimate in (5.16g), which is valid for any global solution u starting from an initial value $u_0 \in U_\varepsilon$ with some appropriate small but positive ε. □

REMARK 1.3. (a) Recall that a pseudo-gradient vector field [**112, 122**] for a C^1 functional $\mathcal{E} : X \to \mathbb{R}$ is a locally continuous map $\mathcal{N} : X \to X$ such that for all $u \in X$ there holds

$$\langle \mathcal{E}'(u), \mathcal{N}(u)\rangle_{X' \times X} \geq \|\mathcal{E}'(u)\|_{X'}$$

and

$$2\|\mathcal{E}'(u)\|_{X'} \geq \|\mathcal{N}(u)\|_X \geq \|\mathcal{E}'(u)\|_{X'}.$$

Note that both Theorem 1.1 and 1.2 apply in particular to those Lipschitz maps that are pseudo-gradient vector fields for the given functional.

(b) In case the C^1 functional $\mathcal{E} : U \to \mathbb{R}$ satisfies the Palais-Smale condition (see the next section for the definition), condition (i) in Theorem 1.2 can be simplified as the requirement that φ is an isolated critical point of $\mathcal{E}$. Note that the latter condition is merely a necessary condition for the uniform asymptotic stability.

2. Convergence and stability of the steepest descent method

We recall [**112**, p. 78] that a C^1 functional $\mathcal{E} : X \to \mathbb{R}$ is said to satisfy the *Palais-Smale* (in short, (*P.-S.*)) condition if every sequence $\{u_n\}_{n\geq 1} \subset X$ such that $\sup_{n\geq 1} |\mathcal{E}(u_n)| < \infty$ and $\|\mathcal{E}'(u_n)\|_{X'} \to 0$ as $n \to \infty$ contains a convergent subsequence.

Although the (P.-S.) condition yields the existence of critical points for a C^1 functional $\mathcal{E}$, it does not in general give the computing procedure for finding the critical points. In practical uses, one goes to the so-called (continuous) *method of steepest descent* which consists of two steps. First, one finds a locally Lipschitz *pseudo-gradient field* $\mathcal{N} : X \to X$ associated to $\mathcal{E}$; this is a map satisfying

$$\begin{cases} \langle \mathcal{N}(u), \mathcal{E}'(u) \rangle_{X\times X'} = \|\mathcal{E}'(u)\|_X^2 = \|\mathcal{N}(u)\|_{X'}^2 \\ \|\mathcal{N}(u)\|_X \leq 2\|\mathcal{E}'(u)\|_{X'} \end{cases} \tag{5.17a}$$

for all $u \in X$. Second, one looks for solutions of the corresponding evolution equation in X :

$$\dot{u}(t) = -\mathcal{N}(u(t)),\ t \geq 0. \tag{5.17b}$$

Relying on the fact that the energy $\mathcal{E}(u(t))$ $(t \geq 0)$ is a nonincreasing function of the time, one hopes that the solutions of (5.17b) enjoy the long-time convergence to one single critical point. However, as shown by Example 3.7 of Chapter 2 (cf. also Palis and de Melo [**93**]), the convergence fails already for some two-dimensional gradient systems. Having known that all solutions to (5.17b) converge, we also have another problem that arises from the practical situation. This is the stability problem of which we ask whether the convergence to one single equilibrium of solutions to (5.17) will be kept when there is some external disturbance coming from an apparatus. More precisely, we ask if solutions to the following equation

$$\dot{u}(t) = -\mathcal{N}(u(t)) + g(t, u(t)),\ t \geq 0 \tag{5.18a}$$

with a "small" perturbation $g(t, u)$ will also converge to some critical point of $\mathcal{E}$ if all solutions to (5.17b) have been known to obey this convergence property.

In order to answer this question, we consider a more general situation where $\mathcal{N} : X \to X$ is a locally Lipschitz map for which there exists a positive constant α such that

$$\text{(5.18b)} \qquad \begin{cases} \langle \mathcal{E}'(u), \mathcal{N}(u) \rangle_{X' \times X} \geq \alpha \cdot \|\mathcal{E}'(u)\|_{X'}^2, \\ \|\mathcal{E}'(u)\|_{X'} \geq \alpha \cdot \|\mathcal{N}(u)\|_X \end{cases} \qquad \forall u \in X.$$

Moreover, the perturbation $g : \mathbb{R}_+ \times X \to X$ is a locally Lipschitz map that satisfies the following smallness condition:

$$\text{(5.18c)} \qquad \begin{cases} \|g(s,u)\|_X \leq g(s) \quad \forall s \geq 0, u \in X; \\ \displaystyle\int_0^\infty \left(\int_t^\infty g(s)^2 \, ds \right)^{1/2} dt < \infty. \end{cases}$$

The growth condition on g of the above type has been used in Chapter 3, see (3.28c). As observed there, there holds

$$\text{(5.18c}'\text{)} \qquad \text{(5.18c)} \Longrightarrow g \in L^1 \cap L^2(\mathbb{R}_+).$$

The importance of gradient inequality of type $\mathcal{G}_k$ in answering the above-mentioned stability problem of the steepest descent method is emphasized in the following theorem.

THEOREM 2.1. *Assume that the C^1 functional $\mathcal{E} : X \to \mathbb{R}$ is bounded below, satisfies the (P.-S.) condition, and satisfies a gradient inequality of type $\mathcal{G}_k$ near every given point of X (with a constant $k \geq 1$ depending on that given point). Then, for each $u_0 \in X$, the Cauchy problem corresponding to (5.18) admits a global solution $u : [0, \infty) \to X$ that converges as $t \to \infty$ to some critical point of $\mathcal{E}$. Moreover, $I(u) = \int_0^\infty \|\dot{u}(t)\|_X \, dt < \infty$.*

PROOF. It is routine to deduce the global existence of the solution u to (5.18a), using condition (5.18b) combined with condition (5.18c). Moreover, it is not difficult to show that any solution of (5.18a) satisfies a growth condition of the form (3.10a,b) (in Chapter 3) under the following choices:

$$(X, Y, \mathcal{E}') = (X, X', \mathcal{E}'),$$

with the map $\gamma := c_1 \cdot \text{Identity}$, and the functions F, G given by

$$F(t) := c_2 \cdot \int_t^\infty g(s)^2 \, ds, \quad G(t) := c_3 \cdot \int_t^\infty g(s) \, ds,$$

where c_1, c_2, c_3 are some appropriate positive constants. Here we omit the details. Note that the convergence $F(t) + G(t) \to 0$ as $t \to \infty$ is ensured by (5.18c$'$).

We want to show the convergence of the solution u by virtue of Theorem 3.3 of Chapter 3 (corresponding to the above choice of spaces and the map γ). To do so, we need to show that the orbit $O(u)$ contains a cluster point $\varphi \in X$ with $\mathcal{E}'(\varphi) = 0$. Once this has been done, then $\mathcal{E}'$ satisfies a gradient inequality of type $\mathcal{G}_k$ near φ, as assumed in the theorem.

To show the existence of the cluster point, we note that

$$\begin{aligned}
-\frac{d}{dt}\mathcal{E}(u(t)) &= \langle \mathcal{E}'(u), -\dot{u}\rangle_{X'\times X} = \langle \mathcal{E}'(u), \mathcal{N}(u) - g(t,u)\rangle_{X'\times X} \\
&\geq \langle \mathcal{E}'(u), \mathcal{N}(u)\rangle_{X'\times X} - \|\mathcal{E}'(u)\|_{X'} \cdot \|g(t,u)\|_X \\
&\geq \alpha \cdot \|\mathcal{E}'(u)\|_{X'}^2 - (\alpha \cdot \|\mathcal{E}'(u)\|_{X'}^2 + g(t)^2/\alpha)/2 \quad \text{(using (5.18b))} \\
&= (\alpha/2) \cdot \|\mathcal{E}'(u)\|_{X'}^2 - (1/\alpha) g(t)^2,
\end{aligned}$$

and thus, by integration,

$$\int_0^\infty \|\mathcal{E}'(u(t))\|_{X'}^2\, dt \leq a + (2/\alpha^2) \cdot \int_0^\infty g(t)^2\, dt < \infty, \tag{5.19}$$

where $a := (2/\alpha)[\mathcal{E}(u_0) - \limsup_{t\to\infty} \mathcal{E}(u(t))] < \infty$ by the assumption that $\mathcal{E}$ is bounded below. Since the function $t \to \|\mathcal{E}'(u(t))\|_{X'}$ is continuous, it is routine to show, using (5.19), that there exists a sequence $t_n \to \infty$ such that $\|\mathcal{E}'(u(t_n))\|_{X'} \to 0$ as $n \to \infty$. Finally, the (P.-S.) condition implies that $\{u(t_n)\}_{n\geq 1}$ contains a cluster point in X, as required. □

As seen in Theorem 3.7 of Chapter 3 as well as the results presented in §4 of Chapter 3 for the Hilbert space setting, once the gradient inequalities involved in the above theorem are of the special weak LS-type, then the growth condition (5.18c) for the perturbation can be relaxed as follows:

$$\int_0^\infty \left(\rho(t) g(t)^2 + \frac{1}{\rho(t)}\right) dt < \infty \tag{5.20}$$

with an increasing and positive function ρ. More precisely, we have the following result.

THEOREM 2.2. *Assume that the C^1 functional $\mathcal{E} : X \to \mathbb{R}$ is bounded below, satisfies the (P.-S.) condition, and satisfies a gradient inequality of weak LS-type near every given point of X. (As seen in Theorem 4.5 of Chapter 2, the validity of such gradient inequalities is true if $\mathcal{E}'$ is analytic and Fredholm.)*

Then, for each $u_0 \in X$, the Cauchy problem corresponding to (5.18a) accompanied by conditions (5.18b) and (5.20) admits a global solution $u : [0,\infty) \to X$ that converges as $t \to \infty$ to some critical point of $\mathcal{E}$. Moreover, $I(u) = \int_0^\infty \|\dot{u}(t)\|_X\, dt < \infty$.

PROOF. The proof is very similar to the one for Theorem 2.1 combined with the one for Theorem 3.7 of Chapter 3. Here we omit the details. □

In many application situations, the steepest descent method is used particularly in the following **optimization problem**:

$$|F(x)|^2 \to \text{Min!} \tag{5.21a}$$

on the finitely dimensional space $\mathbb{R}^n$, where $F : \mathbb{R}^m \to \mathbb{R}^n$ is a continuous map, and $|\cdot|$ denotes the Euclidean norm on $\mathbb{R}^n$. There is a common case, namely, the case where the map F is *real analytic*. In the case that F is real analytic, the function

$$\mathcal{E}(x) := \frac{1}{2}|F(x)|^2 \quad (x \in \mathbb{R}^m) \tag{5.21b}$$

is real analytic as well. Hence, by the classical theorem of Łojasiewicz (Theorem 3.1 (i) of Chapter 2), we conclude that the derivative $\mathcal{E}'$ satisfies a gradient inequality of LS-type near any given point of $\mathbb{R}^m$. Clearly, the function $\mathcal{E}$ is nonnegative and thus bounded from below. Therefore, the above Theorem 2.2 will be applicable if $\mathcal{E}$ furthermore satisfies the (P.-S.) condition.

Since

$$\mathcal{E}'(x)y = \langle F'(x)^* F(x), y\rangle \quad \forall x, y \in \mathbb{R}^m,$$

where $F'(x)$ is the Jacobian matrix of the map F and $F'(x)^*$ is its transpose, we have that

$$\mathcal{E}'(x) = F'(x)^* F(x) \quad \forall x \in \mathbb{R}^m. \tag{5.21c}$$

Thus, the (P.-S.) condition for our present case reads:

Every sequence $\{x_k\} \subset \mathbb{R}^m$ such that $\sup\{|F(x_k)| : k \geq 1\} < \infty$ and

$$|F'(x_k)^* F(x_k)| \to 0 \quad \text{as} \quad k \to \infty$$

is bounded. In particular, if F is coercive in the sense that

$$|F(x)| \to \infty \quad \text{as} \quad |x| \to \infty,$$

then $\mathcal{E}$ satisfies the above-mentioned (P.-S.) condition.

Note that the (P.-S.) condition implies that the function $\mathcal{E}(x)$ attains its absolute minimum at some point of $\mathbb{R}^m$.

Now the steepest descent method corresponds to solving the following Cauchy problem

$$\dot{x} = -F'(x)^* F(x), \quad x(0) = x_0 \in \mathbb{R}^m. \tag{5.22a}$$

Or by taking into account the circumstance perturbations, it corresponds to solving the Cauchy problem

$$\dot{x} = -F'(x)^* F(x) + g(t, x), \quad x(0) = x_0 \in \mathbb{R}^m. \tag{5.22b}$$

Here, we assume that the time-dependent perturbation $g(t, x)$ satisfies the smallness condition of the form (5.20); i.e., there holds $|g(t, x)| \leq g(t)$ for all $(t, x) \in \mathbb{R}_+ \times \mathbb{R}^m$ and there is an increasing and positive function ρ such that (5.20) holds.

Summing up, we have the following result as a consequence of Theorem 2.2 combined with Theorem 1.2 as well as Remark 1.3 (b).

THEOREM 2.3. *Assume that the map $F : \mathbb{R}^m \to \mathbb{R}^n$ is real analytic and satisfies the above (P.-S.) condition.*

(i) Assume the above-mentioned smallness condition for the perturbation $g(t, x)$. Then for any initial value $x_0 \in \mathbb{R}^m$, the Cauchy problem (5.22b)

admits a global solution $x : [0, \infty) \to \mathbb{R}^m$ *that converges as* $t \to \infty$ *to some equilibrium of (5.22a).*

(ii) Let $\varphi \in \mathbb{R}^m$ *be such that* $|F(x)| \geq |F(\varphi)|$ *for all* $x \in \mathbb{R}^m$*. Then* φ *as an equilibrium of system (5.22a) is Lyapunov stable. Moreover,* φ *is uniformly asymptotic stable if and only if* φ *is an isolated equilibrium of (5.22a).*

We point out that in the recent work [**68**] we have shown, for the finite-dimensional case, that gradient inequalities are also sufficient to ensure the convergence to one single critical point of the usual *discrete gradient method.*

3. The structure of equilibria sets of convergent systems

An interesting question that we want to answer follows:

Problem. *If an autonomous system* $\dot{u} = -\mathcal{N}(u)$ *is a convergent system in the sense that every solution of it is a global one and converges as* $t \to \infty$ *to some equilibrium of it, should its equilibria set* $\mathcal{N}^{-1}(0) = \{u \in X : \mathcal{N}(u) = 0\}$ *be very small in some sense?*

The answer to this problem is "no". As shown in the following example, a convergent system is allowed to take any closed subset as its equilibria set.

EXAMPLE 3.1. Take X to be an arbitrary real Banach space that has the property that its dual X' is uniformly convex. Then the functional

$$\mathcal{E}_X(x) := \frac{1}{2}\|x\|_X^2, \quad x \in X \tag{5.23a}$$

has a Fréchet derivative $\mathcal{E}'_X$ coinciding with the *duality map* $\mathcal{F} : X \to X'$ given by

$$\mathcal{F}(x) := \{x' \in X' : x'(x) = \|x\|_X^2 = \|x'\|_{X'}^2\} \quad (x \in X).$$

We refer to Deimling [**32**, Proposition 12.3, p. 115] for a proof of this result. It follows that

$$\|\mathcal{E}'_X(x)\|_{X'} = \|x\|_X \quad \text{and} \quad \langle x, \mathcal{E}'_X(x)\rangle = \|x\|_X^2 \qquad \forall x \in X. \tag{5.23b}$$

Clearly, $\mathcal{E}_X$ attains its global minimum at the origin. Moreover, (5.23b) implies that $\mathcal{E}_X$ satisfies the following optimal and global gradient inequality:

$$\sqrt{2} \cdot (\mathcal{E}_X(x))^{1/2} = \|\mathcal{E}'_X(x)\|_{X'} \quad \forall x \in X. \tag{5.23c}$$

Let K be an arbitrarily given closed set of X containing the origin. Define a function $\rho_K : X \to \mathbb{R}_+$ by

$$\rho_K(x) := \frac{\operatorname{dist}(x, K)}{1 + \operatorname{dist}(x, K)}, \quad x \in X,$$

where $\operatorname{dist}(x, K) := \inf_{y \in K} \|x - y\|$ for each $x \in X$. It is easy to verify that ρ_K is a global Lipschitz function in the sense that

$$|\rho_K(x) - \rho_K(y)| \leq \|x - y\| \quad \forall x, y \in X.$$

This implies that the map $\mathcal{N}_X : X \to X$, given by

$$\mathcal{N}_X(u) := \rho_K(u)u, \quad u \in X,$$

is a locally Lipschitz map. Moreover, we have by (5.23b) that

$$\begin{aligned}\langle \mathcal{E}'_X(u), \mathcal{N}_X(u)\rangle &= \rho_K(u) \cdot \langle \mathcal{E}'_X(u), u\rangle \\ &= \rho_K(u) \cdot \|u\|_X^2 = \|\mathcal{E}'_X(u)\|_{X'} \cdot \|\mathcal{N}_X(u)\|_X.\end{aligned}$$

Therefore, because of the global gradient inequality (5.23c), Theorem 2.1 is applicable for any bounded ball U_σ and thus it yields that the system

$$\dot{u}(t) = -\rho_K(u(t)) \cdot u(t), t \geq 0; u(0) = u_0 \in X$$

is a convergent system taking $\mathcal{N}_X^{-1}(0) = K$ as its equilibria set.

4. Numeric test

We consider the two-dimensional potential

$$V(x,y) \equiv V(r,\theta) := r^2(1-r)^2[2 + \sin(r+\theta)], \tag{5.24}$$

which belongs to the type of potentials treated in Example 3.5 of Chapter 2, under the choice of the functions $A(r) := r^2(1-r)^2$, $B(r) := r$ and $g(s) := 2 + \sin s$. Clearly, V is coercive in the sense that $V(z) \to \infty$ as $|z| \to \infty$. In particular, V satisfies the (P.-S.) condition.

The set of critical points of V is the union of the origin with the unit circle.

As shown in Example 3.5 of Chapter 2, the gradient ∇V satisfies a gradient inequality of LS-type (see (2.0b)), since the above function $A(r)$ is analytic. Thus our Theorem 2.2 applies and yields that all solutions of the following two-dimensional system

$$\dot{z}(t) = -\nabla V(z) + g(t), \quad t \geq 0 \tag{5.25}$$

($z := (x,y) \in \mathbb{R}^2$) are global and bounded, and converge as $t \to \infty$ to some critical point of V, provided that the perturbation g vanishes so quickly that

$$\int_0^\infty t^\tau |g(t)|^2 \, dt < \infty \quad \text{with some} \quad \tau > 1. \tag{5.26}$$

What will happen if g vanishes so slowly that the convergence condition (5.26) fails? We treat this problem numerically. For this purpose, we rewrite (5.25) in its polar coordinates (r,θ) :

$$\begin{cases} \dot{r} = -V_r + g_1(t)\cos\theta + g_2(t)\sin\theta, \\ \dot{\theta} = -\frac{1}{r}V_\theta + \frac{1}{r}(-g_1(t)\sin\theta + g_2(t)\cos\theta). \end{cases} \tag{5.25'}$$

Here $(g_1, g_2) = g$.

We choose perturbations of the type

$$g_\sigma(t) = (15\sin(2t)/(1+t^\sigma), 15\sin(2t)/(1+t^\sigma)),$$

which depend on the parameter $\sigma > 0$. g_σ satisfies (5.26) if and only if $\sigma > 1$. The case $\sigma = 1$ lies just at the threshold of the convergence interval $(1,\infty)$

and thus it could be expected that the solutions corresponding to the case $\sigma = 1$ have a complicated long-time behavior.

We integrate (5.25′) numerically for both cases $\sigma = 1$ and $\sigma = 1.2$. Our numerical result (see Fig. 1 (a) below) corresponding to the case $\sigma = 1$ indicates that the trajectory starting from the point $(1.2, 0) \in \mathbb{R}^2$ does not converge to any critical point V, while for the case $\sigma = 1.2$ (shown in Fig. 1 (b)) the trajectory starting from the same point $(1.2, 0)$ does converge to the critical point $(0, 0) \in \mathbb{R}^2$ of V, as predicted by our theory.

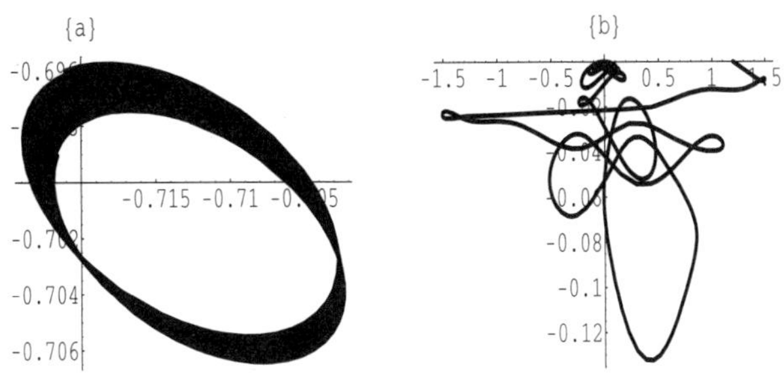

Fig. 1: Plot of trajectories in the xy-plane.
(a) $\sigma = 1$. Nonconvergent trajectory.
(b) $\sigma = 1.2$. Trajectory converging to the equilibrium $(0, 0)$.

Bibliography

[1] R.A. Adams, *Sobolev Spaces*, Academic Press, New York-San Francisco-London, 1975.

[2] S. Aizicovici, E. Feireisl, *Long-time stabilization of solutions to a phase-field model with memory,* J. Evol. Equ. **1** (2001), 69-84.

[3] S. Aizicovici, E. Feireisl, F. Issard-Roch, *Long-time convergence of solutions to the phase-field system,* Math. Methods Appl. Sci. **24** (2000), 277-287.

[4] S. Aizicovici, H. Petzeltová, *Asymptotic behavior of solutions of a conserved phase-field system with memory* (preprint 2002).

[5] L. Alvarez, F. Guichard, P.-L. Lions, J.-M. Morel, *Axioms and fundamental equations of image processing,* Arch. Rational Mech. Anal. **123** (1993), 199-257.

[6] H. Amann, T. Laetsch, *Positive solutions of convex nonlinear eigenvalue problems,* Indiana Univ. Math. J. **23** (1974), 1069-1076.

[7] H. Amann, *Supersolutions, monotone iterations, and stability,* J. Diff. Eqns. **21** (1976), 363-377.

[8] A. Ambrosetti, P.H. Rabinowitz, *Dual variational methods in critical point theory and applications,* J. Funct. Anal. **14** (1973), 349-381.

[9] L. Ambrosio, *Geometric evolution problems, distance function and viscosity solutions,* in *Calculus of Variations and Partial Differential Equations* (Eds. L. Ambrosio, N. Dancer), pp. 5-93, Springer-Verlag, 2000.

[10] B. Andrews, *Monotone quantities and unique limits for evolving convex hypersurfaces,* Intern. Math. Res. Notices **1997**, 1001-1031.

[11] I.S. Aranson, K.A. Gorshkov, A.S. Lomov, M.I. Rabinovich, *Stable particle-like solutions of multidimensional nonlinear fields,* Physica D **43** (1990), 435-453.

[12] G. Aubert, P. Kornprobst, *Mathematical Problems in Image Processing,* Springer-Verlag, 2002.

[13] M.S. Berger, *Nonlinearity and Functional Analysis,* Academic Press, New York-San Francisco-London, 1977.

[14] H. Brezis, *Operateurs maximaux monotone et semigroups de contractions dans les espaces de Hilbert,* Math. Studies **5**, North-Holland Amsterdam, 1973.

[15] G. Caginalp, *An analysis of a phase field model of a free boundary,* Arch. Rational Mech. Anal. **92** (1986), 205-246.

[16] K.C. Chang, *Infinite Dimensional Morse Theory and Multiple Solution Problem,* Prog. Nonlin. Diff. Eq. Appl. **6**, Birkhäuser, 1993.

[17] Z.M. Chen, K.-H. Hoffmann, J. Liang, *On a nonstationary Ginzburg-Landau superconductivity model,* Math. Methods Appl. Sci. **16** (1993), 855-875.

[18] R. Chill, *On the Lojasiewicz-Simon gradient inequality,* J. Funct. Anal. **201** (2003), 572-601.

[19] R. Chill, E. Fašangová, *Convergence to steady states of solutions of semilinear evolutionary integral equations* (preprint 2002).

[20] R. Chill, E. Fašangová, J. Prüss, *Convergence to steady states of solutions of the Cahn-Hilliard equation with dynamic boundary conditions* (preprint 2004).

[21] R. Chill, M.A. Jendoubi, *Convergence to steady states in asymptotically autonomous semilinear evolution equations,* Nonlin. Anal.: Theory, Methods & Applications **53A** (2003), 1017-1039.

[22] S.-N. Chow, J.K. Hale, *Methods of Bifurcation Theory,* Springer-Verlag, 1982.

[23] K. Chueh, C. Conley, J. Smoller, *Positively invariant regions for systems of nonlinear diffusion equations,* Ind. U. Math. J. **26** (1977), 373-392.

[24] P. Constantin, C. Foias, B. Nicolaenko, R. Temam, *Integral Manifolds and Inertial Manifolds for Dissipative Partial Differential Equations,* Springer-Verlag, 1989.

[25] M. Crandall, P.H. Rabinowitz, *Some continuation and variational methods for positive solutions of nonlinear elliptic eigenvalue problems,* Arch. Rational Mech. Anal. **58** (1975), 207-218.

[26] M.C. Cross, P.C. Hohenberg, *Pattern formation outside of equilibrium,* Review of Modern Physics **65** (1993), 851-1112.

[27] B. Dacorogna, *Direct Methods in the Calculus of Variations,* Springer-Verlag, 1989.

[28] E.N. Dancer, *The effect of domain shape on the number of positive solutions of certain nonlinear equations,* J. Diff. Eqns. **74** (1988), 120-156. And **Part II** *in* J. Diff. Eqns. **87** (1990), 316-339.

[29] E.N. Dancer, P. Poláčik, *Realization of vector fields and dynamics of spatially homogeneous parabolic equations,* Mem. Am. Math. Soc. **668** (1999).

[30] E.B. Davies, *Heat Kernels and Spectral Theory,* Cambridge Univ. Press, 1989.

[31] D.G. De Figueiredo, *Lectures on the Ekeland Variational Principle with Applications and Detours,* Tata Institute of Fundamental Research, Bombay, 1989.

[32] K. Deimling, *Nonlinear Functional Analysis,* Springer-Verlag, 1985.

[33] Q. Du, *Global existence and uniqueness of solutions of the time-dependent Ginzburg-Landau model for superconductivity,* Appl. Anal. **53** (1994), 1-18.

[34] K.-J. Engel, R. Nagel, *One-Parameter Semigroups for Linear Evolution Equations,* Springer-Verlag, 2000.

[35] J. Escher, J. Prüss, G. Simonett, *A new approach to the regularity of solutions for parabolic equations* (preprint 2002).

[36] E. Feireisl, *On the long time behavior of solutions to nonlinear diffusion equations on* $\mathbb{R}^n$, Nonlinear. Diff. Eqns. Appl. **4** (1997), 43-60.

[37] E. Feireisl, F. Issard-Roch, H. Petzeltová, *Long-time behaviour and convergence towards equilibria for a conserved phase-field model* (preprint 2002).

[38] E. Feireisl, F. Issard-Roch, H. Petzeltová, *A non-smooth version of the Lojasiewicz-Simon theorem with applications to non-local phase-field systems* (preprint 2003).

[39] E. Feireisl, F. Simondon, *Convergence for degenerate parabolic equations,* J. Diff. Eqns. **152** (1999), 439-466.

[40] E. Feireisl, F. Simondon, *Convergence for semilinear degenerate parabolic equations in several space dimensions,* J. Dynamics Diff. Eqns. **12** (2000), 647-673.

[41] E. Feireisl, P. Takáč, *Long-time stabilization of solutions to the Ginzburg-Landau equations of superconductivity,* Monatsh. Math. **133** (2001), 197-221.

[42] J. Fleckinger-Pellé, H.G. Kaper, P. Takáč, *Dynamics of the Ginzburg-Landau equations of superconductivity,* Nonlin. Anal.: Theory, Methods & Applications **32** (5) (1998), 647-665.

[43] M. Giaquinta, G. Modica, J. Souček, *Cartesian Currents in the Calculus of Variations II,* Springer-Verlag, 1998.

[44] B. Gidas, W.-M. Ni, L. Nirenberg, *Symmetry and related properties via the maximum principle,* Comm. Math. Phys. **68** (1979), 209-243.

[45] D. Gilbarg, N.S. Trudinger, *Elliptic Partial Differential Equations of Second Order,* Springer-Verlag, 1983.

[46] V.L. Ginzburg, L.D. Landau, *On the theory of superconductivity,* Zh. Eksp. Teor. Fiz. (USSR) 20, (1950) 1064-1082 (Engl. transl. *in* D. Haar, L.D. Landau; Men of Physics, Vol. I, Pergamon Press, Oxford, 1965, pp. 138-167).

[47] L. Gross, *Logarithmic Sobolev inequalities and contractivity properties of semigroups,* In "Dirichlet Forms" (eds. Fabes *et al.*), pp. 54-88, Springer-Verlag, 1993.

[48] C. Gui, W.-M. Ni, X. Wang, *On the stability and instability of positive steady states of a semilinear heat equation*, Comm. Pure Appl. Math. **45** (1992), 1153-1181.

[49] C. Gui, W.-M. Ni, X. Wang, *Further study on a semilinear heat equation*, J. Diff. Eqns. **169** (2001), 588-613.

[50] P.F. Guang, G.F. Wang, *A fully nonlinear conformal flow on locally conformally flat manifolds,* J. reine angew. Math. **557** (2003), 219-238.

[51] J.K. Hale, *Asymptotic Behavior of Dissipative Systems,* Math. Surveys and Monographs, Vol. 25. Amer. Math. Soc., Providence, RI, 1988.

[52] J.K. Hale, P. Massatt, *Asymptotic behavior of gradient-like systems,* in *Dynamical Systems* Vol II (Eds. A.R. Bednarek, L. Cesari), Academic Press, Inc., New York, 1982, 85-101.

[53] J.K. Hale, G. Raugel, *Convergence in gradient-like systems with applications to PDE,* Z. Angew. Math. Phys. **43** (1992), 63-124.

[54] A. Haraux, *Exponentially stable positive solutions to a forced semilinear parabolic equation*, Asymptotic Analysis **7** (1993), 3-13.

[55] A. Haraux, *A hyperbolic variant of Simon's convergence theorem,* in *6th Conference on Evolution Equations and Their Applications in Life Sciences, Bad Herrenalb (1998),* Eds. G. Lumer, L. Weis. Lecture Notes in Pure and Applied Mathematics, vol. **215**, Marcel Dekker, New York, 2000, pp. 253-263.

[56] A. Haraux, *Positively homogeneous functions and the Lojasiewicz gradient inequality* (preprint 2002).

[57] A. Haraux, *Slow and fast decay of solutions to some second order evolution equations* (preprint 2003).

[58] A. Haraux, M. A. Jendoubi, *Convergence of solutions of second-order gradient-like systems with analytic nonlinearities,* J. Diff. Eqns. **144** (1998), 313-320.

[59] A. Haraux, M. A. Jendoubi, *Convergence of bounded weak solutions of the wave equations with dissipation and analytic nonlinearity,* Carl. Var. and PDE **9** (1999), 95-124.

[60] A. Haraux, M. A. Jendoubi, *Decay estimates to equilibrium for some evolution equations with an analytic nonlinearity,* Asymptotic Analysis **26** (2001), 21-36.

[61] A. Haraux, M. A. Jendoubi, *On the convergence of global and bounded solution of some evolution equations* (preprint 2002).

[62] A. Haraux, P. Poláčik, *Convergence to a positive equilibrium for some nonlinear evolution equations in a ball,* Acta Math. Univ. Comenian **2** (1992), 129-141.

[63] D. Henry, *Geometric Theory of Semilinear Parabolic Equations,* Springer-Verlag, 1981.

[64] S.-Z. Huang, *Constructing Navier-Stokes equations with attractors of arbitrary complexity,* Physics Letters A **282** (2001), 1-8.

[65] S.-Z. Huang, *Constructing universal pattern formation processes governed by reaction-diffusion systems,* Electron. J. Diff. Eqns. **2002** (2002), 1-12.

[66] S.-Z. Huang, *Inequalities for submarkovian operators and submarkovian semigroups,* Math. Nachr. **243** (2002), 75-91.

[67] S.-Z. Huang, *Existence of compact attractors for semilinear heat semiflows with rapidly growing nonlinearity* (preprint 2003).

[68] S.-Z. Huang, *Convergence to one single limit of discrete gradient methods ensured by gradient inequalities* (preprint 2005).

[69] S.-Z. Huang, P. Takáč, *Global smooth solutions of the complex Ginzburg-Landau equations and their dynamical properties,* Discrete and Continuous Dynamical Systems **5** (1999), 824-848.

[70] S.-Z. Huang, P. Takáč, *Convergence in gradient-like systems which are asymptotically autonomous and analytic,* Nonlin. Anal.: Theory, Methods & Applications **46** (2001), 675-698.

[71] M.A. Jendoubi, *Convergence of global and bounded solutions of the wave equation with linear dissipation and analytic nonlinearity,* J. Diff. Eqns. **144** (1998), 302-312.

[72] M.A. Jendoubi, *A simple unified approach of some convergence theorem of L. Simon,* J. Funct. Anal. **153** (1998), 187-202.

[73] M.A. Jendoubi, *Exponential stability of positive solutions to some nonlinear heat equations,* Portugaliae Math. **55** (1998), 401-409.

[74] Y. Katznelson, *An Introduction to Harmonic Analysis,* 2nd Ed., Dover Publ., Inc., New York, 1976.

[75] V. Khatskevich, D. Shoiykhet, *Differentiable Operators and Nonlinear Equations,* Birkhäuser Verlag, 1994.

[76] K. Kurdyka, T. Mostowski, A. Parusiński, *Proof of the gradient conjecture of R. Thom,* Ann. Math. **152** (2000), 763-792.

[77] O.A. Ladyženskaja, V.A. Solonnikov, N.N. Ural'ceva, *Linear and Quasi-linear Equations of Parabolic Type,* Amer. Math. Soc. (Providence, R. I.), 1968.

[78] F.-H. Lin, Q. Du, *Ginzburg-Landau vortices: dynamics, pinning, and hysteresis,* SIAM J. Math. Anal. **28** (1997), 1265-1293.

[79] P.-L. Lions, *Structure of the set of the steady-state and asymptotic behavior of semilinear heat equations,* J. Diff. Eqns. **53** (1984), 362-386.

[80] V.A. Liskevich, Yu.A. Semenov, *Some inequalities for submarkovian generators and their applications to perturbation theory,* Proc. Amer. Math. Soc. **119** (1993), 1171-1177.

[81] S. Łojasiewicz, *Une propriété topologique des sous-ensembles analytiques réels,* Colloques internationaux du C.N.R.S. # 117. Les équations aux dérivées partielles, 1963.

[82] S. Łojasiewicz, *Ensembles semi-analytiques,* I.H.E.S. Notes, 1965.

[83] S. Łojasiewicz, *Sur la geometrie semi- et sous-analytique,* Ann. Inst. Fourier (Grenoble) **43** (1993), 1575-1595.

[84] Z.-M. Ma, M. Röckner, *Introduction to the Theory of (Non-Symmetric) Dirichlet Forms,* Springer-Verlag, 1992.

[85] J. Malý, W.P. Ziemer, *Fine Regularity of Solutions of Elliptic Partial Differential Equations, Mathematical Surveys and Monographs* **51** (1997), Amer. Math. Soc., 1997.

[86] H. Matano, *Convergence of solutions of one-dimensional semilinear heat equation,* J. Math. Tokyo Univ. **18** (1978), 221-227.

[87] V.G. Maz'ja, *Sobolev Spaces,* Springer-Verlag, 1985.

[88] K. Nagasaki, T. Suzuki, *Radial and nonradial solutions for the nonlinear eigenvalue problem $\Delta u + \lambda e^u = 0$ on annuli in $\mathbb{R}^2$,* J. Diff. Eqns. **87** (1990), 144-168.

[89] R. Nagel (Ed.), *One-parameter Semigroups of Positive Operators,* Springer-Verlag, 1986.

[90] R. Nagel, J. Voigt, *On inequalities for symmetric submarkovian operators,* Arch. Math. **67** (1996), 308-311.

[91] J.W. Neuberger, *Sobolev Gradients and Differential Equations,* Springer-Verlag, 1997.

[92] A. Nowel, Z. Szafraniec, *On trajectories of analytic gradient fields,* J. Diff. Eqns. **184** (2002), 215-223.

[93] J. Palis, W. de Melo, *Geometric Theory of Dynamical Systems, an Introduction,* Springer-Verlag, 1982.

[94] A. Pazy, *Semigroups of Linear Operators and Applications to Partial Differential Equations,* Springer-Verlag, 1983.

[95] P. Poláčik, K.P. Rybakowski, *Nonconvergent bounded trajectories in semilinear heat equations,* J. Diff. Eqns. **124** (1996), 472-494.

[96] P. Poláčik, F. Simondon, *Nonconvergent bounded solutions of semilinear heat equations on arbitrary domains,* J. Diff. Eqns. **186** (2002), 586-610.

[97] P. Poláčik, E. Yanagida, *On bounded and unbounded global solutions of a supercritical semilinear heat equation*, Math. Ann. **327** (2003), 745-771.

[98] Y. Pomeau, P. Manneville, *Wavelength selection in cellular flows,* Phys. Lett. **75A** (1980), 296-298.

[99] J. Prüss, *Evolutionary Integral Equations and Applications*, Birkhäuser Verlag (1993).

[100] G. Raugel, *Dynamics of partial differential equations on thin domains,* in *Lecture Notes on Dynamical Systems*, Vol. 1609, pp. 208-315, 1994.

[101] P. Rybka, K.-H. Hoffmann, *Convergence of solutions to the equation of quasi-static approximation of viscoelasticity with capillarity,* J. Math. Anal. Appl. **226** (1998), 61-81.

[102] P. Rybka, K.-H. Hoffmann, *Convergence of solutions to the Cahn-Hilliard equation,* Comm. PDEs **24** (5 & 6) (1999), 1055-1077.

[103] F. Santz, *Non oscillating solutions of analytic gradient vector fields*, Ann. Inst. Fourier (Grenoble) **48** (1998), 1045-1067.

[104] J.T. Schwartz, *Nonlinear Functional Analysis,* Gordon and Breach, 1969.

[105] R.E. Showalter, *Monotone Operators in Banach Spaces and Nonlinear Partial Differential Equations,* Mathematical surveys and monographs vol. **49**, Amer. Math. Soc. (RI), 1997.

[106] B. Simon, *Schrödinger semigroups,* Bull. (New Series) of AMS **7** (1982), 447-526.

[107] L. Simon, *Asymptotics for a class of non-linear evolution equations, with applications to geometric problems,* Ann. Math. **118** (1983), 525-571.

[108] L. Simon, *Theorems on Regularity and Singularity of Energy Minimizing Maps,* Lectures in Math. ETH Zürich. Birkhäuser Verlag, 1996.

[109] Hal L. Smith, *Monotone Dynamical Systems: An Introduction to the Theory of Competitive and Cooperative Systems,* AMS (Providence, R. I.), 1995.

[110] J. Smoller, *Shock Waves and Reaction-Diffusion Equations,* 2nd Ed. Springer-Verlag, 1983.

[111] E.M. Stein, *Topic in Harmonic Analysis Related to the Littlewood-Paley Theory,* Princeton Univ. Press, 1970.

[112] M. Struwe, *Variational Methods,* 2nd Ed., Springer-Verlag, 1996.

[113] M. Struwe, *Geometric evolution problems,* in *Nonlinear Partial Differential Equations in Differential Geometry* (Eds. R. Hardt, M. Wolf), pp. 259-339, Amer. Math. Soc., 1996.

[114] J.B. Swift, P.C. Hohenberg, *Hydrodynamics fluctuations at the convective instability,* Phys. Rev. A **15** (1977), 319-329.

[115] P. Takáč, *Stabilization of positive solutions for analytic gradient-like systems,* Discrete and Continuous Dynamical Systems **6** (2000), 947-974.

[116] Q. Tang, *On an evolutionary system of Ginzburg-Landau equations with fixed total magnetic flux,* Comm. Partial Diff. Eqns. **20** (1995), 1-36.

[117] Q. Tang, S. Wang, *Time dependent Ginzburg-Landau equations of superconductivity,* Physica D **88** (1995), 139-166.

[118] R. Temam, *Infinite-Dimensional Dynamical Systems in Mechanics and Physics,* Springer-Verlag, 1988.

[119] M. Tinkham, *Introduction to Superconductivity,* 2nd Edition. McGraw-Hill, Inc., New York, 1966.

[120] M.F. Bidaut-Veron, L. Veron, *Nonlinear elliptic equations on compact Riemannian manifold and asymptotics of Emden equations,* Invent. Math. **106** (1991), 489-539, with an erratum published in Invent. Math. **112** (1993), 445.

[121] L. Veron, *Conformal asymptotics of the isothermal gas spheres equation,* in *Nonlinear Differential Equations and Their Equilibrium States, 3* (Eds. N.G. Lloyd, W.M. Ni, L.A. Peletier and J. Serrin). Birkhäuser Verlag, 1992.

[122] M. Willem, *Minimax Theorem,* Birkhäuser Verlag, 1996.

[123] K. Yosida, *Functional Analysis,* Spinger-Verlag, 1980.

[124] T.J. Zelenyak, *Stabilization of solutions of boundary value problems for a second-order parabolic equation with one space variable,* Differntsial'eye Uravneniya **4** (1968), 17-22.

[125] X.Q. Zhao, *Dynamical Systems in Population Biology,* Springer-Verlag, 2003.

[126] W.P. Ziemer, *Weakly Differentiable Functions,* Springer-Verlag, 1989.

Index